Stacked Decks

Stacked Decks

*Building Inspectors and the
Reproduction of Urban Inequality*

ROBIN BARTRAM

THE UNIVERSITY OF CHICAGO PRESS   CHICAGO AND LONDON

The University of Chicago Press, Chicago 60637
The University of Chicago Press, Ltd., London
© 2022 by The University of Chicago
All rights reserved. No part of this book may be used or reproduced in any manner whatsoever without written permission, except in the case of brief quotations in critical articles and reviews. For more information, contact the University of Chicago Press, 1427 E. 60th St., Chicago, IL 60637.
Published 2022
Printed in the United States of America

31 30 29 28 27 26 25 24 23 22   1 2 3 4 5

ISBN-13: 978-0-226-81906-8 (cloth)
ISBN-13: 978-0-226-82114-6 (paper)
ISBN-13: 978-0-226-82113-9 (e-book)
DOI: https://doi.org/10.7208/chicago/9780226821139.001.0001

Library of Congress Cataloging-in-Publication Data

Names: Bartram, Robin, author.
Title: Stacked decks : building inspectors and the reproduction of urban inequality / Robin Bartram.
Other titles: Building inspectors and the reproduction of urban inequality
Description: Chicago ; Illinois : The University of Chicago Press, 2022. | Includes bibliographical references and index.
Identifiers: LCCN 2021054319 | ISBN 9780226819068 (cloth) | ISBN 9780226821146 (paperback) | ISBN 9780226821139 (ebook)
Subjects: LCSH: Building inspectors—Illinois—Chicago. | Building inspection—Social aspects—Illinois—Chicago. | Housing—Illinois—Chicago. | Equality—Illinois—Chicago. | Discrimination in housing—Illinois—Chicago.
Classification: LCC HD7304.C4 B37 2022 | DDC 363.509773/11—dc23/eng/20211206
LC record available at https://lccn.loc.gov/2021054319

Contents

Preface   vii

Introduction   1

CHAPTER 1.   Stacked Decks   22

CHAPTER 2.   Building Inspections   48

CHAPTER 3.   Rentals and Relative Assessments   75

CHAPTER 4.   Helping Out Homeowners: Changing Faces and Stubborn Realities   99

CHAPTER 5.   Justice Blockers   128

CONCLUSION.   Reshuffling the Deck   148

Acknowledgments   163

APPENDIX A.   Methodology   165

APPENDIX B.   Building Violation Counts   179

APPENDIX C.   Map of Strategic Task Force Inspections   183

Notes   185

References   203

Index   225

Preface

I did not like the view from my desk. The window looked out onto two dumpsters that were constantly overflowing in the alley behind our building. Bag after bag piled up until they toppled and cascaded down the sides, splitting and spitting their household waste insides onto the ground. There, they turned rancid or were trodden on, driven over, pecked at by pigeons, or delighted in by rats. It wasn't just household waste. Bedbug-infested mattresses, broken cabinets, and other discarded furniture joined the trash bags. And, just as soon as the refuse collection had whisked it all away—every Monday and Thursday in our neighborhood of Chicago—the dumpsters seemed to be already full again and ready to overflow. It didn't take me long to get sick and tired of this revolting pattern.

I decided to act. The dumpsters I could see belonged to the building across from ours. I took down the name of the management company from the sign on the front of the building, looked them up online, and called to complain. I called three times. Twice my calls went straight to voicemail. The other time, I spoke to someone who assured me he would tell the janitor. But nothing changed. The dumpsters continued to overflow, and I continued to be annoyed. I picked up the phone to call the City. But I hesitated. Would this get the owner in trouble? I wasn't sure I wanted that. Wasn't it the tenants—and not the owner—who were causing the problem? The building wasn't in great shape: there were cracks in bricks over a doorway, the paint that used to be blue was faded, and basement windows were broken. Maybe these were signs that the building owner didn't have money for upkeep and was struggling financially? I didn't want to make matters worse. Yet again, wasn't it the owner's responsibility to make sure this mess got cleaned up, to provide more dumpsters, or to schedule frequent enough garbage collections? I decided to wait before calling the City. I'd do some research first.

I looked up the building's management company online. Reviews from angry tenants lamented mold, leaks, mice, broken elevators, and unanswered requests to fix issues. I also asked my own building manager what she knew about the building. She told me that the owners were "slumlords." They lived in the suburbs and owned a handful of big apartment buildings in the city, which were all poorly kept. I had the information I needed. The owners, I deduced, did not care about their tenants but continued to rake in rent payments. I reckoned negligent property owners who collected rents from multiple apartment buildings deserved whatever the City would throw at them. I'd show them! My call might take them down a peg or two and teach them a lesson! I called the City.

My decision to make the call hinged on my interpretation of this building. I had made an assessment about the owners—without ever seeing or talking to them—based on pieces of information about the building, including its material condition. This was not a one-off. We do this all the time. We often drive or walk down a street and make decisions about who lives there, without seeing anyone. Many of us decide not to walk down some streets because they look dangerous, or to consider moving to a block because it looks peaceful, safe, or hip. We use buildings and other material aspects of the urban environment as clues to the kinds of people that live behind closed doors. I had seen faded paint, broken windows, and crumbling brickwork and decided this meant something about the finances and character of the building owners. This makes sense because we all do it, but it's remarkable if you really think about it. Why should faded paint mean anything to us? Aren't there all kinds of reasons why paint might be faded? What are we really noticing when we decide a block looks dangerous, safe, or hip? Accurate or not, our eyes communicate something to our minds, and tell us whether a building or a neighborhood looks neglected, poor, wealthy, dangerous, or safe. What our eyes tell us matters. We use interpretations of buildings to make decisions about people, and these decisions can have severe consequences for urban inequality.

Introduction

It was bitterly cold, even by Chicago standards, and one tenant had had enough. She lodged this online request with the City about the three-unit rental building where she lived:[1]

> No heat for three weeks!!! The heat has not been on in the building for three weeks!!! It is very cold in here. I can see my breath in my room. I use my stove for heat at night and I am scared that I might start fire. Other naybors doing the same thing. Manager said building owner don't want to pay for more heat this season. March and April and sometimes May still get cold outside!!!!!!! I pay rent to avoid homeless! I pay for heat!!! This is wrong! Pease help us!!!

In response, a city building inspector evaluated the building in question—a historic but run-down greystone in an ethnically diverse neighborhood. The inspector cited the building for numerous violations of the building code and brought the property owner to building court. But in court, the inspector asked the judge to be lenient because the building owner could barely afford to pay for water and property taxes. Forcing the owner to fix other issues might jeopardize the owner's ability to provide water for his tenants. Sure enough, the judge asked that the owner fix the heating but allowed him more time to find funds before fixing up other issues with the property. The court case lasted five years, as the owner continued to come up short. Eventually, the building was sold, and the tenant who made the call had to move out. But inspection records show that the issues in the building persisted. Six years later, residents still used ovens to heat their units, pulling up chairs and sitting in front of the open oven doors on cold days. Ovens make for inefficient and dangerous heating devices—using them as such can double utility bills and cause fires.

Three blocks away, an anonymous citizen—probably a neighbor—put a call in to the City about a small, shabby-looking condo building: "Bricks falling out from top—this is a 4-unit condo. Bricks are out all the way around. Pieces of cement have fallen in alley. Decaying. Also back steps are shaky." The building was wedged between a large historic single-family home and a newish three-story condo building. A building inspection ensued, and, on hearing that the condo board was almost bankrupt, the inspector recommended that the owners tap into a program—a partnership between the city government and a nonprofit corporation—that covers up-front costs of emergency repairs. In contrast to the previous case of the greystone owner's unsuccessful five-year search for funds, the program ensured that the repair work got done quickly—the masonry issues have been resolved, and the building is now safe. This was not a win-win for the owners, however, because they must still find a way to repay the program for the pricey repair work. It will be difficult for any of the owners to sell their units or get a new mortgage until the debt is paid. What's more, paying off debts like this can mean property owners don't have enough money to pay their mortgages or property taxes. This catch-22 situation leads to foreclosures—when banks or other lenders take possession of buildings if the owners cannot make timely payments.

Across the street, a few days later, another neighborhood resident had reached the end of his patience with his landlord. He picked up the phone and called the City about his twenty-plus–unit rental building:

> None of the doors in the building lock . . . a lot of rats and roaches and the landlord doesn't do anything about it. Water leaking. Mold growing all over the place bothering asthmatics in building. Overwhelming smell of dead rats.

This building was in bad shape. A sweep of the building by a team of building inspectors, performed in response to several complaints that had been received, found broken windows, dangerous wiring, sagging exterior walls, and collapsing door frames to complement the rats, roaches, leaks, and mold. The inspection report also lists a cracked lavatory and improper size toilet tank. The Buildings Department had come down hard in response to the landlord's negligence. The sixty-plus building code violations uncovered by the meticulous inspection brought the building owner to building court, and the judge did not show leniency in this instance, insisting that the owner make fast progress on repairs. All the issues have now been fixed. The building was gut renovated—totally overhauled—in the process. Leaking pipes have been replaced with brand-new bathroom fixtures and chrome appliances. Fresh paint covers the clusters of small black and green spots from mold that once marred walls throughout the building. There are also other changes: the landlord now charges much higher rents and tenants, like the one who made the call, have been priced out. The conditions of the building may have improved, but to whose benefit?

Only a few blocks separate these three buildings, but they introduce a variety of goings-on in the city: from slumlordism, poverty, and dangerous housing conditions, to displacement, government housing programs, and upscaling. Yet these buildings and their inhabitants meet a similar—and all too familiar—fate: tenants face the double jeopardy of dire housing conditions and displacement, and homeowners with scant resources remain at risk of losing their homes. What's more, they meet this fate despite efforts to assist low-income residents and punish negligent landlords. This is the paradox at the heart of this book: that attempts to mete out justice do little to challenge injustice. We see city workers—building inspectors in this case—going easy on some property owners and coming down hard on others. I call these actions *stabs at justice* because they are motivated by a sense of injustice or unfairness and because they are intended to level the playing field, in the short run at least. As the fates of these buildings reveal, stabs at justice do not always go to plan. A lack of regulation amid a for-profit housing market, for example, means that landlords pass financial penalties on to tenants, while acts of compassion toward low-income homeowners are not enough to avoid costly outcomes. Features of the housing market block stabs at justice. I use the term *justice blockers* to depict the multitude of obstacles that prevent stabs at justice from really challenging injustice in a way that destabilizes it. It is because of justice blockers that stabs at justice end up exacerbating unjust and unequal situations. Those who are already marginalized in urban housing markets remain so.

Social scientists use terms like "the reproduction of inequality" or "structural inequality" to capture the persistence of injustice. But when we listen to those on the ground who navigate the uneven places and systems that we theorize, we learn a lot more about how inequality actually works. Building inspectors, police officers, welfare workers, bus drivers, nurses, insurance agents, landlords, judges, and hundreds of others all work within and up against the unequal social world. And they possess sophisticated yet colloquial ways of making sense of and talking about inequality—often without using the word inequality. I suggest that we learn from the voices of those in these positions and adopt the term *stacked deck* to capture what they describe and what we might otherwise generalize as inequality. I posit this term in this book, but it exists on the ground as a frame. Erving Goffman defines frames as "schemata of interpretation" that enable people to perceive, identify, label, and thereby organize their experiences in the world.[2] Stacked decks—and ideas about who decks are stacked by, for, and against—are ways of understanding the unequal positions that people occupy in the social world.

The stacked deck appropriately describes the frames of people who navigate unequal places and situations because they tend to highlight relationships of inequality, using adages such as "the poor stay poor and the rich get richer" and connect the plight of the "little guy" to the success of "the man." Police officers, for example, draw sharp distinctions between themselves as good guys and others as bad guys or assholes to validate the work they do.[3] The stacked deck is thus a relational concept, by which I mean that it describes the relationship between the haves and the have-nots. The same decisions, laws, and policies create poverty *and* wealth, marginality *and* privilege.[4] Like the physical "decks" that adjoin houses and that are stacked upon each other in multi-unit buildings, stacked decks comprise hierarchies and relational connections between people, places, and property. They are constructed. They are intended to be sturdy and are built to last. But they can also collapse. Decks on buildings require maintenance to ensure that they do not break under the weight of people who gather on them. Stacked decks also require maintenance; they are actively stacked and maintained by people, not just shuffled by chance or by the invisible hand.[5] Governments and citizens make explicit decisions that create enduring patterns in poverty, wealth, marginality, and privilege, especially along lines of race, class, and gender.

The stacked deck analogy also makes clear that disparity comprises various compounding blocks stacked on top of one another: government legislation layered atop economic policy, for example, or a new technological or bureaucratic tool tacked on to some century-old legislation, or a housing program that intersects with educational policy. Identifying this characteristic of stacked decks also gives us insight into how we might undo and remake them, piece by piece.[6] One way is to listen to and learn from the people who navigate them, even if we disagree with their ideas and decisions. We have to understand how things are broken to begin to fix them. But first we need to know more about stacked decks and how they are constructed.

How the Deck Is Stacked

Contemporary society is full of stacked decks. And sociologists have depicted many of them. The stacked deck manifests clearly in Gwendolyn Purifoye's ethnography of public transit in Chicago,[7] where we see Chicagoans of color routinely waiting for trains and buses amid the fumes from bus engines, odors from overflowing dumpsters, and the flies the dumpsters attract. These are the features of the transit center on the majority Black South Side of the city. Riders boarding trains at the other end of the line—in a diverse North Side neighborhood—are met with sights and smells of restaurants and coffee shops. This is not by chance, but due to racialized policies about housing, transit, and land use, for example, that have dictated what gets built where. Stacked decks in educational settings produce obvious disparities in school classrooms and facilities. Carla Shedd illustrates these disparities through the eyes of high school students in Chicago.[8] While White students get to use world-class athletic facilities and walk to class down well-lit hallways, majority Black schools contend with dilapidated gyms with broken equipment and hallways with paint peeling from the walls.

Police forces and courts that under-protect, over-police, and disproportionately incarcerate communities of color are features of the stacked decks of criminal justice in which White boys are five times less likely to get arrested than Black boys.[9] As a result, White boys avoid fines and criminal records, incarceration, and the harsh reality of re-entry. Workplaces in which women earn less and in which their contributions are underappreciated are also stacked decks.[10] Auditing policies within the IRS produce a stacked deck of taxation, meaning workers on cotton farms in the Mississippi Delta pay more tax than the US president.[11] And regressive property tax assessments mean that owners of low-priced properties end up in complicated systems of tax delinquency and are forced out of their homes.[12] In short, society is full of stacked decks that overlap and amass to form vast disparities. Disparities in turn beget disparities, most frequently along lines of race, class, and gender. The stacked deck is both cause and manifestation of disparity.

Housing is a clear example of a stacked deck. The deck comprises multiple overlapping pieces: from government policies to local practices. Housing disparities have been created, for example, by government mortgage insurance policy requirements established by the Federal Housing Administration (which we now recognize as redlining), exclusionary zoning that propped up property prices in White neighborhoods, and fearmongering tactics on the part of real estate agents that precipitated White flight. The toxic combination of racist decisions about who could get a decent mortgage and who was targeted for a risky loan has increased the racial wealth gap. Meanwhile, the US tax system rewards owners of expensive properties and concentrates wealth, while lobbying ensures minimal protections for tenants in rental housing. These policies and practices, which we will learn more about in chapter 2, overlap, and create a stacked deck that manifests in dramatically unequal housing—from blocks of vacant properties and empty lots to luxury downtown lofts and gated mansions—in cities across the US.

People stack the deck. Often these people are politicians, policymakers, CEOs, bank managers, government officials, and lobbyists. In the context of urban inequalities, we also know that landlords, lenders, realtors, and residents can stack the deck for themselves and against others. Housing policy is set up to concentrate wealth on purpose by the people who have the most to gain. In other contexts, school boards, parents, judges, and prosecutors stack decks. People have created policies and institutionalized practices—from the War on Drugs, Don't Ask Don't Tell, and broken-windows policing to profit-oriented lowball insurance payouts and teaching to the test—that create other stacked decks.

The stacked deck is relational.[13] Poverty and wealth are not the result of a person's attributes; they are "actively produced through unequal relationships between the financially secure and insecure."[14] Urban America provides a multitude of examples. Banks stack the deck for the financial benefit of investors and against homeowners with subprime mortgages, for example. The precarity of the urban poor is a consequence of the power of others, such as landlords, developers, and the police.[15] Similarly, White wealthy suburbia is not separate from poor urban communities of color. Rather, in Keeanga-Yamahtta Taylor's words, these residential landscapes are part of "a single U.S. housing market," "defined by its racially discriminatory, tiered access," with "each tier reinforcing and legitimizing the other."[16] Governments actively opted to leave Black urban neighborhoods to deteriorate and zoned them to permit industry, taverns, liquor stores, nightclubs, and houses of prostitution. They did so to spare White neighborhoods and suburbs these conditions and their detrimental effects on property prices.[17] There is rarely ever profit without precarity, nor precarity without profit. In short, the deck is stacked against one person or neighborhood because it is stacked in favor of another.

But we're not very used to thinking about inequality in this way. For example, Rebecca Solnit recalls an NPR story during the COVID-19 pandemic that revealed that, in heterosexual households, "women's careers [were being] crushed . . . because they're doing the lioness's share of domestic labor."[18] Solnit points to the language used to frame this issue. Men were "written out of the story altogether," she noticed, as if the additional labor that had "landed on the shoulders of women . . . had fallen from the sky rather than been shoved there by spouses." Solnit continues: "I have yet to see an article about a man's career that's flourishing because he's dumped on his wife." As Solnit makes clear, we too easily forget both sides of the story—the relational aspect of stacked decks.

The same person can have the deck stacked for and against her in different contexts. Categories of profit and precarity within the stacked deck can be sliding scales and moving targets. A landlord, for example, contributes to a deck stacked against low-income renters when she raises rents in her building, but she may also struggle to make ends meet as property taxes or maintenance costs increase over the years. She occupies various positions within the stacked deck of the housing market. The same is true in other contexts too. As Mary Pattillo argues, in spite of the generations of discrimination that have disadvantaged Black neighborhoods, there are advantages to living in majority Black areas—from lower mortality rates to increased emotional well-being for Black people.[19] Similarly, in the face of fewer resources, Black children attain more education and are more likely to go to college than White students of similar backgrounds, and HBCUs are positively correlated with positive educational outcomes, successful careers, and high wages.[20] Pattillo urges us to acknowledge the advantages of being Black alongside everything we know about the way Black neighborhoods and Black schools are disadvantaged by the stacked deck.[21] We can all be subjects and objects of inequality within multiple hierarchies,[22] and this can make inequality murky and hard to pinpoint.

Positions are contingent, but some people are more consistently advantaged and disadvantaged by the stacked deck. Banks, lenders, developers, and White middle-class homeowners, on average, fare better than low-income renters, single mothers, or homeowners of color, for example. There is relative consistency in who the deck is stacked for, by, and against. This is significant because it helps us to identify categories. Categories may obscure variation, but they are useful tools that help people to make sense of the world, organize, and make claims. Categories can motivate.

And because the stacked deck produces categories, this means that the stacked deck can be a call to action.

We are most used to hearing about categorizations shaping actions in the context of stereotypes. Social scientists and media reports point to the deleterious consequences of cognitive biases and stereotypes, in which people make both explicit, intentional decisions and off-the-cuff choices that penalize and harm those in marginalized positions. Many of these accounts come from criminological and sociological research on policing and perceptions of disorder.[23] A host of research shows that people fall back on stereotypes when faced with uncertainty. But people can be motivated by something other than stereotypes. More than we realize, people are motivated by stacked decks.[24]

How Stacked Decks Motivate

People see, read about, hear of, or witness manifestations of stacked decks every day. The morning news features a report about racial inequities in access to healthcare, for example, focusing on high rates of asthma in a Latinx community. Opening a newspaper reveals a story about how the gender pay gap stifles women's efforts to pay off student debt, followed by a story about a bank's racially disparate lending record. But not everyone sees the same stacked deck. A jogger might run past historic mansions every morning, and not see this wealth as part and parcel of a broad pattern of inequality. Maybe two people on the same bus notice once-shabby buildings getting facelifts, but they make different judgments. While one sees investment and beautification, the other perhaps worries about the existing residents being priced out. And a building inspector may have a different view of the stacked deck than, say, a teacher. Meanwhile, I might be angered by the rent my landlord charges because I do not know how much she pays in overhead each month. We do not all see the stacked deck in the same way, and our views do not always align with the reality of inequality.[25] How one person sees the stacked deck has as much to do with the beholder as with material conditions, and is shaped by overlapping factors: our social location, our recourses and clout, and our interactions with stacked decks.

People occupy different *social locations*, by which I mean they have different identities and experiences based on their intersecting demographic and social positions, such as income, gender, race, region, or nativity. It is not a perfect science, but sociologists theorize, for example, that women have different social locations than men. As such, a male and female roommate have different social locations, even if they are both Latinx, college-educated teachers in their thirties. Similarly, we assume a Black factory worker has a different social location than a Black university professor, by virtue of their occupational status, which we relate to class. Crucially, we expect that these different social locations will produce distinct ways of seeing the world. Despite their similarities, we assume that the roommates perceive the world differently from one another—and that, irrespective of their common racial identity, the factory worker has a different view of the world than the professor. Broadly speaking, sociologists trace people's opinions and worldviews to sociodemographic differences, most commonly differences in class, race and ethnicity, and gender.[26] People with different social locations perceive inequality in different ways. And people with similar social locations perceive it in similar ways. We do not all see the same stacked deck.

Our perceptions depend a lot on what we think we have the power to do. People possess—and perceive themselves as possessing—various amounts of professional or political *recourses and clout*.[27] A city politician, for example, might feel empowered to impact urban inequities, whereas her chauffeur might not. But this politician may also be constrained by aspects of her work. Perhaps she is concerned about securing reelection or raising campaign funds, and thus refuses to diverge from the status quo. The chauffeur, on the other hand, may have more freedom to act, despite her relative lack of clout. A building inspector may feel stuck between a rock and a hard place when he inspects a building that is unsafe but is home to a family with nowhere else to go. A lack of recourse can mean people feel stuck and powerless. For example, as Amanda Lewis and John Diamond show, even though teachers, parents, and students know schools treat Black and White students unequally, they feel like there is not much they can do to change it.[28] No one can identify a suitable course of action, so nothing changes. Our previous experiences also shape how empowered we feel. A resident who called 911 about a suspicious person in her neighborhood may feel empowered if her call was met with a swift and calm response from responding police officers. And this empowerment may motivate her to get involved in community events. She may have felt disempowered, however, if her call elicited no response, or if the ensuing police presence escalated the situation. These latter situations may incline the resident to shy away from her community and avoid calling 911 in the future.

Our *interactions and experiences* with stacked decks also shape how we discern inequality. Things we see in person are especially powerful. A bus ride into a city center takes us past designer stores and offers glimpses of their wealthy customers. Meanwhile, a drive to work might take us past a police officer pulling over a Black driver. We might see the signs of a recent eviction—heaps of furniture and clothing strewn outside a house—when out walking a dog. These sights, scenes, and situations motivate some people to act because they are material cues and, as such, convey with particular power different aspects of the stacked deck.[29] Glimpses of customers in expensive-looking clothing confirm that designer stores are places of wealth and exclusion that many cannot afford, and in which many would not feel welcome. A traffic stop of a Black driver confirms what we know about racial disparities in policing. And the heaps of clothes and furniture remind us of recent coverage of eviction crises in US cities. Each of these sights aligns with categories—along lines of race, wealth, and poverty—that make sense because of our understanding of contemporary disparity. As such, each sight may evoke the stacked deck. Our specific experiences matter, too. Someone who regularly drives through blocks of foreclosed homes and empty lots may think differently about housing disparities than someone whose journey to work takes them through leafy streets with lavish houses and manicured yards. As existing literature suggests, seeing both ends of the spectrum has a dramatic effect on how people think about disparity,[30] and particular places, such as neighborhoods and schools, can be especially powerful mechanisms that illuminate the cause of disparities.[31] Importantly, what we see and experience—as well as who we are and what we have the capacity to do—shapes how we look at the world and the extent to which we are motivated by stacked decks.

Efforts to Address Injustice and How They Often Fail

Stacked decks can motivate people to mete out justice. But let's take a step back for a minute, because people use the word "justice" all the time without specifying what it means. It can be used to mean fairness, but colloquially it also often carries the connotation of retribution. Definitions of justice are also related to what we think should happen for justice to be achieved, such as retribution (punishment for injustice) or reparation (compensation for injustice), for example. Whether punitive or reparative, very often people's attempts to mete out justice are not long-term, well-thought-out actions. Instead, they are what I call *stabs at justice*.

To take a stab at justice is to take aim at immediate and small-scale goals. Stabs at justice are rarely as planned or as calculated as what we generally think of as "resistance" or "contestation." While they stem from normative ideals, they are less likely to be geared toward wholesale change. Instead, stabs at justice are efforts to make things fairer in the moment at hand. They are, borrowing from James Scott,[32] actions that are located within the vast territory between overt collective defiance and complete compliance. As Scott argues, people do what they can, with the resources and tools at their disposal. Stabs at justice also tend to stem from evaluations of the manifestation of the stacked deck that we witness, rather than evaluations of the stacked deck as a whole. In other words, stabs at justice are motivated not by objective measures of inequality, but by what an individual perceives as unfair or just. Not only is it hard to imagine that people carry around knowledge of objective measures of inequality, but also, as Wendy Bottero makes clear, we all have different ideas of what justice looks like and who deserves it.[33]

White newcomers to communities of color who are aware of gentrification and want to diminish the impact of their presence—social preservationists, in the words of Japonica Brown-Saracino[34]—might make stabs at justice by attempting to steer the sale of a house on their block. They may be warm and welcoming to house hunters of color and offer them information about the neighborhood, in the hopes they can encourage them to move in. By contrast, they might ignore a White couple who visit the house, thereby appeasing their own sense of culpability in neighborhood change. A crossing guard can decide when to turn his sign, whom to let across the road, and who must wait for a long line of cars to pass. He might make a stab at justice by stopping traffic to usher a child across whose parent's night shift means he is regularly late for school. And a landlord might allow a single mother to make a late rent payment because she is aware of the obstacles facing single mothers in the workplace, but might not show the same generosity to a young professional couple with two incomes. When a perception of how the deck is stacked motivates actions like these, they are stabs at justice. These examples also reveal that stabs at justice can be simultaneously effective and ineffective. The social preservationist may succeed in getting a new neighbor of color, but this has no impact on the wealth gap that fuels racial disparities in who, on average, can afford to move into neighborhoods as house prices increase. The crossing guard's compassion might save the child from one day of tardiness, but this has no bearing on the undesirable hours many parents are compelled to work. The landlord provides real material relief for the single mother, but her stab at justice does nothing to prevent the tenant from facing the same issues next month, and thus does not impact the stacked deck.

Stacked decks motivate stabs at justice, but they also obstruct action. This book offers the concept of *justice blockers* to demonstrate that the exacerbation of inequality is neither wholly intended nor unintended. In fact, although this distinction has long been a topic of academic interest,[35] it is not particularly useful. Rather, justice blockers chaperone stabs at justice and render them tools of the stacked deck. Justice blockers can be people, laws and policies, technological innovations, financial instruments, material objects, or even networks and resources. Celeste Watkins-Hayes, for example, demonstrates that changes to welfare policy coupled with a lack of professional development cause inconsistency in welfare workers' evaluations, and mean their stabs at justice put welfare recipients at risk of ineligibility.[36] In this case, welfare policy and training procedures are justice blockers. Newly installed cameras may cause a bus driver to lose her job because she was caught letting some people ride for free. As a result, no bus drivers allow anyone to ride for free—no matter their circumstances. The people whose marginalization prompts stabs at justice pay the price. Justice blockers—from welfare policy to surveillance cameras—prevent justice from being achieved.

It works the other way around too. A police officer may opt to crack down on white-collar crime to make amends for her role in the overpolicing of petty crime, for example. But after arrests have been made, white-collar criminals have the connections to bargain their way out of convictions and can provide the police with names of their street-level associates, who then end up with jail time. Coming down hard on people with power often reverberates in the direction of those with less power. Building code inspectors may punish wealthy landlords for leaking faucets and bathroom mold, but the landlords' wealth also enables them to renovate their units as they make repairs and even generate more wealth from rent hikes justified by shiny new bathrooms.

Although stabs at justice do not lead to predetermined outcomes, they are channeled and exacerbated by their institutional contexts.[37] In her ethnography of mobile home parks, Esther Sullivan demonstrates that state protection in Florida had the effect of disempowering residents in the eviction process and aggravating their trauma.[38] In this case, the implementation of state policy undermines its protectionist objective. It is a justice blocker. When a judge in eviction court orders an eviction without considering a tenant's claim about substandard housing conditions, she is acting as a justice blocker.[39] When city council members fail to regulate "ban the box" policies that prohibit asking job applicants about their arrest and conviction record, they are making it possible for justice to be blocked. Poverty also frequently blocks stabs at justice—for example, by limiting the ability of first-generation students to benefit from college scholarships because they also have to work to help support family members or be able to afford rent. The concept of justice blockers thus helps us to understand how the stacked deck persists in spite of stabs at justice. The poor may stay poor and the rich might get richer or escape justice, but not only because of the actions of individuals. Rather, it is because of justice blockers that make the stacked deck stubborn. We can only successfully challenge inequality if we address justice blockers.

Taken together, these concepts—stacked decks, stabs at justice, and justice blockers—equip us to understand the persistence of disparities, and illuminate opportunities for change. Revealing how stabs at justice are rendered futile helps shine a light on where we should focus our efforts to reshuffle the deck. While assessments of how the deck is stacked motivate action, justice blockers frustrate and obstruct these efforts, and they will continue to do so unless we put them front and center of our efforts to make change. But identifying justice blockers requires that we pay close attention to the experiences of people who work within and try to circumvent the stacked deck. We need to hear from them and walk in their shoes to understand what they are up against.

Stacked Decks and Justice in Frontline Urban Governance

There are at least two reasons why frontline urban governance is a useful context to see how stacked decks motivate stabs at justice and how justice blockers obstruct justice. First, the people at the front line of urban governance—building inspectors, medics, and police officers, for example—spend their workdays driving through city streets and meeting urban residents. Characterized by size, density, and heterogeneity, cities often make disparities evident.[40] Second, as a condition of their positions as government employees at the interface of the state and the public, those at the front line of urban governance have discretion, wield some power, and need to categorize complex situations.[41] Discretion is arguably the most important characteristic of frontline work. Heavy caseloads, limited time and resources, and complex cases require frontline workers to make on-the-spot decisions about who has a legitimate claim or complaint, whether to believe someone, when to overlook a mistake, or when to come down hard to send a message.[42] They rely on the stacked deck for these tasks.

This book introduces us to the world of building inspectors in Chicago. In doing so, it builds on recent studies that put faces on frontline workers—the otherwise invisible laborers who populate city streets and government offices.[43] The book also emphasizes the importance of seemingly mundane aspects of urban life. The labor of building inspectors is even more invisible than that of many other frontline workers because of its mundanity. Inspectors don't engage in high-speed chases, make arrests, rescue cats from trees, or resuscitate people on the roadside. Even though the average American spends 87 percent of their time inside buildings,[44] we rarely worry about whether our roofs will collapse or whether our electrical wiring was installed correctly. Building inspectors are employed to ensure these things. Whether you know it or not, a building inspector has probably been to the place you live—when it was being built or perhaps more recently.

While some building inspectors focus on new construction and permits or specialize in plumbing or heating and ventilation systems, the focus of this book is the daily work lives of the building inspectors who spend their time responding to calls from the public about housing conditions. Calls and complaints about building conditions come in every day to cities across the US, concerning rental units, owner-occupied properties, and vacant buildings. In response, city governments send out building inspectors to evaluate whether the issues that motivate the calls amount to violations of the building code. The building code covers a lot of ground: from overgrown weeds in a yard, leaking ceilings, flaking paint on windows, rats, and roaches, to hazardous and partially collapsed roofs and walls. Non-permitted basement conversions, rental buildings without smoke detectors, and bedbugs also fall under the purview of building codes, as do many more issues you will encounter in this book.

The building code is enforced unevenly. This is the case for all laws and regulations, but it is particularly important that we understand how this applies for building code enforcement because of the life-and-death consequences of compliance and violations. Buildings collapse and catch fire.[45] But building code enforcement in Chicago is uneven in a way that sociologists might struggle to make sense of using existing accounts of frontline work and inequality, because inspectors do not help the rich get richer or the poor stay poor—at least not in the way we might expect. Existing studies of frontline work focus on the paternalistic and disciplinary logics of governance that disproportionately discipline and sanction low-income residents of color, for example.[46] Some reveal that discipline is contingent on local politics and organizational structures.[47] Others posit the importance of the social identities of frontline workers themselves, showing that frontline workers might be more lenient toward people of the same race or ethnicity as themselves.[48]

None of these logics explain the decisions that inspectors make. Inspectors are motivated by injustice and try to even the playing field for property owners in the city, but not because they share racial or ethnic identities with those to whom they show compassion. Inspectors are a group of mostly White working-class men who grew up in White ethnic enclaves famous for racism. Yet they make efforts to help out low- and moderate-income property owners and landlords in communities of color who struggle to maintain their buildings. Inspectors show compassion toward the owners of small rental buildings like the one with no heat I mentioned at the start of this chapter. They also show compassion to low-income homeowners like the condo owners dealing with falling bricks. What's more, they try to punish negligent landlords, development companies, politicians, banks, and mortgage and insurance companies. They come down hard on big rental buildings managed by professional landlords like the one with leaks, rats, and roaches.

The decisions that inspectors make are particularly surprising when we consider that they work for a city renowned for corruption, racial profiling, and residential segregation. A wealth of prior studies would cause us to expect inspectors to align with the growth machine or to pad their own wallets: to protect profit-seeking investors in the lucrative real estate market, prioritize the exchange value of property, or take bribes to smooth things over for wealthy developers. Research on policing, broken windows, and disorder would lead us to expect that inspectors would disproportionately police low-income and minority neighborhoods and perceive more issues in these places than in Whiter and wealthier areas. Instead, inspectors interpret the housing landscapes that uneven development and investment create, and they wield their interpretations of these landscapes to guide their assessments about who deserves leniency or punishment, and how to allocate justice.[49]

If we look further, we can see that inspectors' decisions about leniency and punishment are attempts at doling out moments of reparative and retributive justice. There is ideological sentiment behind their actions. They enact retributive justice toward professional landlords (i.e., they make them "pay"), for example, and they enact a version of reparative justice toward struggling homeowners (i.e., they try not to add to their headaches). While their versions of retribution and reparation are short-lived and often just "stabs" in the direction of justice, this is the recourse available to inspectors.

Building inspectors offer a window into how people can be motivated by disparity because they drive through and assess the city's unequal housing landscape every day. Their work makes it clear that the deck is stacked favorably for some and unfavorably for others. People live close to one another, but worlds apart in terms of resources, well-being, and opportunities. Foreclosed homes with boarded-up windows are only minutes away from shiny new developments. Streetscapes change dramatically from one block to the next. Sometimes the lack of proximate disparity is just as revealing—empty lot after empty lot with no interruption, or row upon row of wealth concentrated in historic red brick townhomes. Both visible difference and sameness tell stories of disparity.

Chicago's disparate housing landscape is a physical manifestation of a stacked deck. And this motivates building inspectors. It is not only the state of the buildings, or the plight of their inhabitants, that matters to inspectors. What buildings communicate to inspectors about wealth, poverty, profit, and precarity are just as important when they make evaluations about building code violations and decide whether to cite a property, send it to court, add it to the city's demolition schedule, recommend it for city initiatives, or overlook issues. Moreover, inspectors do not categorize buildings as singular cases. They slot buildings into categories based on typologies of relationships between people and places, and on relational positions within the unequal city. While every building has issues, not every issue leads to a citation or a court case. Nor do inspectors respond in the same way to the same violation in different contexts. Furthermore, inspectors may not insist that building owners correct violations, and may even go out of their way to assist them with alternatives. Inspectors have to decide whom to punish and when to show leniency. The stacked deck helps them make these decisions by providing a lens through which to categorize people, property, and places.

Just like other frontline workers, inspectors' perception of the stacked deck stems from their social location, the recourses available to them and their personal or professional clout, and their experiences in the city (we learn more about this in chapter 3).[50] Inspectors show benevolence to people against whom they perceive the deck to be stacked, and punish those the deck favors. Inspectors rail against negligence and profit they deem unfair. They try to protect property owners in communities of color, help out small-time landlords, and steer the benefits of city initiatives. In this way, inspectors express an awareness of exploitation along the lines of many social scientists and political progressives. But while these camps understand exploitation in a relational way—that problematizes both the undeserved profit of the wealthy and the immiseration of the poor—inspectors tend only to act upon one side of the relational dynamic when it comes to tenants. Inspectors overlook the precarity of tenants, in large part because they have little legal recourse to help them. Furthermore, inspectors underestimate the extent to which the deck is stacked in favor of Whites. I call this a "White blind spot." White blind spots obscure the relational nature of racial inequality. This problem exists beyond building inspections in part because ordinary White wealth is often not as visible as poverty. The form injustice takes matters, because if we cannot see it, it can be hard for us to know it exists. The way the stacked deck is manifested obstructs attempts to change it.

Inspectors' stabs at justice almost always get turned on their head. Think back to the buildings at the start of this chapter. Despite inspectors' efforts, the tenant who made the call about the three-unit rental building with no heat was evicted when the building sold. The current owner appears to have done little to repair issues in the building; last winter tenants were still heating their frigid units with their kitchen ovens. The condo owners who were implicated in the second call promptly fixed the issues with the masonry, after a building inspector recommended that they participate in a city initiative. But as a result, they are now in debt and risk losing their homes and investments if they cannot pay. Finally, the tenants who lived with the roaches, rats, and mold in the large rental building found themselves priced out when their landlord raised rents after making court-mandated renovations. These scenarios exemplify a broader pattern across the city that reproduces housing inequality in spite of inspectors' attempts to blunt it.

Although inspectors try to shape the city according to their perceptions of the stacked deck, the options for action available within the context of frontline urban governance are limited and work to direct and stymie stabs at justice. Justice blockers—such as features of the contemporary housing market, characteristics of the legal system, and the material manifestation of racialized poverty—chaperone inspectors' actions and render their stabs at justice tools of injustice, reproducing the kind of disparities that together make up the stacked deck. This does not, however, mean we should ignore stabs at justice.

Unless we understand stabs at justice, actions may appear to directly reproduce inequality, when in fact inequality is often reproduced in spite of rather than directly because of actions. This is a critical distinction. By attending to how inequality persists despite efforts to change it, this book challenges common understandings of the social world. Conventionally, inequality is understood to persist because of intentional actions or unintended consequences. Theories that center intentionality (or the lack thereof) miss the point and do not equip us to understand how inequality persists. I show, instead, that consequences are rendered inevitable by the structures in which frontline work is embedded. The word "rendered" is important here. The reproduction of inequality is not inevitable, but contingent on justice blockers. Thus, it is *rendered* inevitable and can be rendered un-inevitable too. By upending how we think inequality endures, we can shine a new light on opportunities for change. We need to fully understand how the stacked deck endures if we are to find a way to reshuffle the deck. Doing so means considering how people's positions within the stacked deck make them susceptible to justice blockers. As Iris Young argues, attempts at achieving justice will not create equality without explicit attention to power and domination, and to the distinct needs of those in different social locations.[51]

In short, cities are stacked decks. They are sites of vast disparities in racial wealth, health, education, and well-being. But this stacked deck also motivates. As we will see in this book, disparities in the city inspire and organize action. Built environments—and the inequity they embody—motivate building code inspectors. But features of this unequal world also hinder the actions they inspire and work as justice blockers. This tension—between motivation and obstruction—makes inequality particularly stubborn and hard to change. This is a story of how the stacked deck gets reproduced even when people are trying to do the opposite.

The Project

This book draws on a mixed-methods project that pairs fieldwork and interviews with content analysis of the building code and statistical analyses of almost a million records of building violations in Chicago. My two years of fieldwork began in 2015, when I worked as an intern at the Buildings Department and conducted interviews and ride-alongs with inspectors. Ride-alongs spanned Chicago: from the blocks of bungalow homes on the city's far Northwest Side, to lakeside redevelopment projects, and the disinvested and partially desolate landscape of the South and West Sides, complete with their pockets of new investment and historic strongholds of durable institutions and communities. I also observed building violation cases in building court, where inspectors attend as expert witnesses and help decide which violations get resolved and which buildings get demolished or saved. My interviews with building court judges and attorneys helped me to understand the legal framework of building code enforcement.

Interspersed throughout the book are the results of my statistical analyses of building violations data, 311 requests, and housing market data, which clarify the relationships between code enforcement and trends in inequality, illuminating correlations between Chicago's housing conditions, race, class, tenure, and the uneven enforcement of code violations over time and across the city. Additional FOIA-requested data and content analysis of the municipal building code enabled me to piece together a picture of code enforcement in action, the bureaucratic and legal landscape in which code enforcement is embedded, and the ramifications of code enforcement for people and property in the city. I supplemented these city-wide data with a focused study of one neighborhood in Chicago to investigate code enforcement at the local level. Here I observed community meetings and events, and conducted interviews with local politicians, landlords, and property owners. More details about my methods and data can be found at the end of the book.

Overview of the Book

To orient readers to Chicago's disparate housing landscape, chapter 1 takes us on a guided tour of the city, detailing the histories and legacies and contemporary policies and practices that have created such a starkly stacked deck. We visit foreclosed homes on the city's Southwest Side, a single-family home on the North Side that is a conduit for intergenerational wealth, a luxury high-rise rental building on the South Side lakefront with a history of disrepair, and smaller rental buildings turned condos in a rapidly changing neighborhood on the Northwest Side. The material conditions of each of these buildings and their environs are relational—that is, the wealth and luxury in one building or neighborhood is the result of the poverty and dilapidation in another. Each depicts a different aspect of the stacked deck, and they are all products of historic and contemporary contexts, at the level of the neighborhood, the city, and national politics.

Chapter 2 begins with a puzzle. Calls concerned tenants and neighbors make to the City about housing conditions mirror Chicago's geography of race and wealth. Yet the volume of calls does not map perfectly onto the distribution of the violations that inspectors dole out. So what do inspectors do when they turn up at a building? What prompts an inspector to record a violation? Why do their reports not mirror calls made by the public about building conditions? How much discretion do inspectors have? Are they corrupt? Chapter 2 provides answers to these questions and many more. We discover exactly how the stacked deck motivates inspectors' decisions about building code violations. Specifically, we learn that their social location, their recourses and clout, and their experiences with inequality cause them to use their discretion in patterned and very intentional ways. They choose between leniency and punishment in every decision they make, and they use the stacked deck as a guide.

The next two chapters recast leniency and punishment as stabs at justice. In chapter 3, we take an in-depth look at inspectors' stabs at justice in the context of the rental market. We meet inspector Eddie as he inspects a run-down rental property in a neighborhood notorious among inspectors and the public for poverty and vacant lots. We hear Eddie weighing up the violations, before deciding to punish the landlord and praying that she go to "landlord hell." A deep dive into the judgment calls inspectors like Eddie make every day reveals distinctions inspectors make among rental properties: between big, professional companies, slumlords, corner-cutters, small-time landlords and family buildings, and deserving and undeserving tenants. Distinctions are not always accurate, and categories sometimes obscure variation. But witnessing inspectors make and justify these distinctions sheds light on the relative nature of the evaluative tools inspectors use to categorize people and property in the stacked deck. Evaluations are not just about one person or building or neighborhood, but also about its relationship to others. This chapter posits that we should pay attention to how all kinds of people use relative assessments—based on relationships within stacked decks—to make decisions.

In chapter 4, we turn to inspections of owner-occupied property. This chapter illuminates decisions like inspector Danny's as he opts to not write up a dilapidated single-family home in a Latinx neighborhood. Features of the stacked deck, including precarious ties to homeownership, evoke inspectors' compassion. We also join inspectors in changing neighborhoods and see how these contexts prompt leniency for newcomers as well as old-timers. We see inspectors using stabs at justice to even out the benefits of public-private initiatives and to steer property owners away from precarity. We see how moments of crisis—like the housing crash—change and expand inspectors' categorizations of profit and precarity. But not all categorizations and distinctions of people and property change. Some are persistent and stubborn. As we see, the obstinacy of racial disparities in housing conditions obstructs stabs at justice in communities of color and makes it very hard for inspectors to make a difference. Examining how inspectors make stabs at justice in the owner-occupied housing market reveals racialized patterns in the obduracy of the stacked deck, thereby also shining a lens on what needs to happen to reshuffle the deck.

Chapter 5 provides a framework to understand how housing inequality is produced in spite of inspectors' actions. We revisit some properties from previous chapters and find out what happened after the inspectors made their stabs at justice. Statistical analyses of housing and rental market data allow me to trace justice blockers' stymieing effects on stabs at justice. We also spend time in building court to better understand how legal recourse—or the lack thereof—obstructs inspectors' attempts at leniency. The chapter offers the concept of justice blockers to enable us to better understand the reproduction of inequality. Importantly, the chapter illuminates that the reproduction of inequality is not inevitable, but provisional on justice blockers.

The conclusion revisits and summarizes the book's key argument—that stacked decks motivate and obstruct stabs at justice—and expands the conceptual apparatus to other areas. Academics and practitioners need to grasp interplays between disparities and motivations, I argue, to illuminate ways to navigate justice blockers and to successfully enact justice. Doing so could mean different outcomes for the three buildings I described at the beginning of this introduction. If small-time landlords could access funds to help with repairs, and if there were regulations on rent increases, for example, renters in both buildings could avoid displacement and live in decent conditions. If governments revamped initiatives to help homeowners fix up their properties and eschewed private interests, the condo owners could avoid foreclosure. Using the lessons learned from building inspectors navigating the stacked deck, this chapter provides broad suggestions for how we might adapt the housing market and legal system to enable rather than obstruct justice, and to reshuffle the deck.

CHAPTER ONE

Stacked Decks

In card games, a stacked deck means that the chances of winning are not equal because players have uneven hands. Maybe someone has all the aces, or five consecutive spades. Depending on the game you are playing, either of these hands could guarantee you a win. Stacked decks can be a result of a dealer manipulating a pack of cards to ensure that someone gets a good hand. They can also occur when players have uneven chances to win—for example, if a professional poker player plays a novice. Not all stacked decks are zero-sum games in which the winner takes all. There are gradations of good and bad hands that up peoples' chances of winning and losing. But whatever their gradation, these hands are relational, meaning that one is related to and dependent on another. In a game where all the cards are dealt, the strength of your hand determines the strength of someone else's.

Cities are stacked decks: vastly uneven places where extreme wealth and poverty go hand in hand. The following pages take us on a tour of one city—Chicago—and its disparate housing conditions. We travel through a landscape of vacant lots, foreclosed homes, and institutional neglect, before visiting streets of concentrated wealth. We see pockets of luxury and profit amid neighborhood poverty, and visit a community dealing with displacement and upscaling. The tour takes in some familiar historical and contemporary practices that have stacked the deck for some residents at the expense of others: from redlining and contract sales, to predatory lending, uneven development, property takings, financialization, evictions, and foreclosure. These actions have allowed wealth extraction, slumlordism, and displacement, which stockpile wealth and exacerbate poverty. Thanks to sociologists and historians, we already know a good amount about why areas of concentrated poverty look the way they do, and how the deck has been stacked against their residents. But we know much less about the underlying conditions that shape the look and feel of wealthier neighborhoods. Although wealth and poverty seem worlds apart, they are relational, meaning that some homes and neighborhoods are dilapidated because resources are channeled to and hoarded within others.[1] This chapter uses a tour of Chicago to explain the underlying relational reasons for disparities in housing conditions and in the issues that building inspectors face.

The Geography of Chicago

Chicago's geography exhibits stark disparity. Chicagoans divide their city rhetorically into the North, South, and West Sides, which are depicted in the map in figure 1. Chicago's sides are more than just names; they refer to and invoke huge differences in racial and class demographics. We can see this in the maps in figures 2 and 3, which depict median household income and the White (non-Hispanic) population in Chicago, respectively.[2] The North Side houses most of the city's White population, while the South and West Sides are (today) predominately Black. The North Side is also wealthier than the South and West Sides. The Northwest and Southwest Sides are more heterogeneous, with large Latinx populations. The labels Chicagoans use obscure heterogeneity, but by and large they denote segregation.

Chicago is segregated by race, income, and wealth, and this means different parts of the city look different.[3] The South and Southwest Sides have streets of empty lots and disrepair, while North Side housing stock is denser and is rarely allowed to decay. The "side" terminology is handy because it conveys critical material differences. It is especially useful for social scientists, many of whom deal in averages. The terminology is also meaningful; Chicagoans—on the North, South, and West Sides—use these words.

But we should not forget that averages can conceal outliers, variation, and places that do not fit the mold. The general understanding of Chicago's South Side as poor and neglected, for example, conceals the area's rich architecture and culture and strong institutions, and the pride, joy, and durability of many South Siders and their communities.[4] It also obscures the fact that, as Mary Pattillo has made clear, there are middle-class and upper-middle-class Black people living in Chicago.[5]

FIGURE 1. Geography of Chicago. Map by Richard Campanella.

Similarly, labeling the North Side as wealthy can ignore the slum-like conditions in some North Side apartments. Moreover, referring to Chicago in terms of sides can reify them, and this can be stigmatizing. The ease with which we make these distinctions can also mean we forget that segregation is not natural or inevitable. As Natalie Moore points out, segregation is such a well-known characteristic of Chicago that we often

STACKED DECKS   25

shrug it off or sweep it under the rug.⁶ But, like all stacked decks, segregation has a history, and it does not persist without maintenance. Our tour of Chicago begins by revealing some of the origins of Chicago's segregated stacked deck and how it manifests and endures in the city's housing.

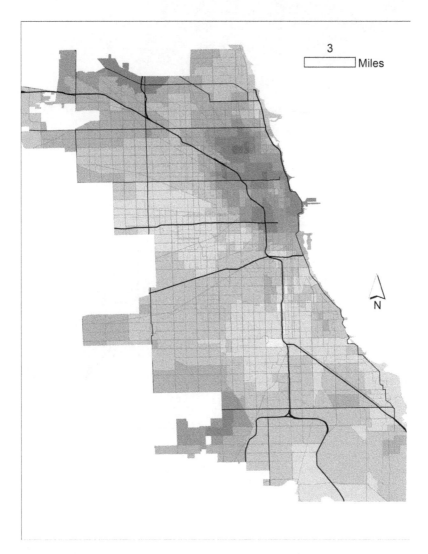

FIGURE 2. Median household income in Chicago, 2010. Shadings show median household income at the census tract level, with lowest incomes in lighter shades and highest incomes in darker shades. Map by Richard Campanella.

FIGURE 3. White population in Chicago, 2012. Shadings show percentage White population at the census tract level, with lower percentages of White residents in lighter shades and higher percentages in darker shades. Map by Richard Campanella.

Institutional Neglect on the Southwest Side

The first stop on the tour is a block of South Loomis Boulevard, in a Southwest Side neighborhood. Many of the buildings on this block are vacant and have accrued handfuls of building code violations over the years. This is an inspector's report for one house on the block:

> front wall at 2nd open mortar joints
> repair exterior platform
> front concrete treads open mortar joints
> front at exterior door, rear at basement door gaps around door
> rear at basement wall open mortar joints, buckling
> 1st living room windows don't hold open
> all elevations rotten window frames
> int wall/ceiling peeling paint
> 1st living room, 1st foyer, 2nd floor kitchen walls and ceiling peeling paint and mold at 2nd
> 1st rear kitchen broken pane.
> int wall/ceiling chip plaster
> 1st kitchen ceiling holes in plaster
> 1st kitchen floor torn tiles holes around radiator on floor
> —Building inspector report for a house on South Loomis Boulevard, 2007

The mold, holes, and buckling mortar in this report troubled the house on the corner of South Loomis for many years. The owners could not afford to fix them. But the house is empty now, the owners long gone. In place of a welcome mat, an orange Buildings Department sign is pasted on the front door declaring that the vacant building is awaiting demolition. These signs are common in this part of Chicago, but vacancy and demolition are not the only stories that buildings on South Loomis tell. Across the street, a two-flat recently changed ownership. Its previous owner had lived in the building for decades, renting out the second unit to an extended family member (what many Chicagoans call a "family building"). Maintenance issues and unpaid property taxes piled up, however, and the longtime owner had no choice but to sell. The new owner is an investment company based in New Jersey that rents the property out. The company fixed up some severe issues, but did not tend to the yard, which is overgrown and littered. This building and its overgrown yard sit across the street from an empty lot where a single-family home stood until it was demolished recently. Grass and weeds are just starting to grow up through the dirt. Between them, these three addresses have accrued over sixty building violations in ten years, from faulty wiring, rotting wooden siding, and buckling brickwork to unstable porches and roofs, and demolition orders. These are listed online, if you know where to look, as is information on whether issues have been addressed. Most on Loomis have not. They linger online like a building's version of a criminal record, seeming to speak volumes about the condition of the buildings but nothing about the underlying causes of their deterioration.

Black Belt, White Flight, and Redlining

Until the 1960s, this Southwest Side neighborhood was majority White ethnic—mostly first- and second-generation European immigrants. It was close to the area of Chicago known as the "Black Belt," which was home to most of the city's Black population—the majority of whom had migrated from the South in the first half of the twentieth century. The Black Belt was getting increasingly crowded, but African Americans had few other options for housing due to pervasive discrimination. Meanwhile, conservative backlash against the New Deal ushered in government programs that demolished what affordable housing was left in the name of slum clearance. Elsewhere in the city, White Chicagoans went to great— and violent—lengths to prevent African Americans from moving beyond the borders of the Black Belt.[7] As in other White neighborhoods, homeowners in this Southwest Side neighborhood used racial covenants that prohibited property owners from selling or renting to African Americans.[8] White Chicagoans also bombed Black homes as part of their campaigns of terror to scare African Americans away from White neighborhoods.

But racial demographics did change in many South and West Side Chicago neighborhoods. A combination of racism and panic over institutional neglect and declining property values (stoked by the real estate industry) led to a broad residential shift. This phenomenon of Whites leaving neighborhoods as African Americans moved in is commonly known as "White flight."[9] The hard-won arrival of the newcomers was costly, in part due to the practice that we now call "redlining," in which the Federal Housing Administration used maps that marked communities of color as unstable and thus too risky for mortgage insurance.[10] Black families were denied mortgages and were left with few options other than contract sales, which—as Beryl Satter details[11]—promised (but did not guarantee) legal ownership only after years of hefty payments. Black families in Southwest Side neighborhoods—and elsewhere—paid higher monthly payments toward their contract sale than they would have paid on a conventional mortgage. A clear extraction of wealth, these payments amounted to an average 84 percent markup over the life of the contract.[12] Contract sellers became millionaires.[13]

Contract sales are one example of the exploitation—on the part of property owners, landlords, and real estate agents—of Black Chicagoans' lack of residential options. Exploitation also meant that prices actually initially increased in the very neighborhoods that Whites were abandoning for fear of declining property values. Yet the average Black resident did not benefit. Despite the high prices Black residents paid to landlords, their homes frequently had no heat or electricity, but plenty of rodents, rotten wood, and broken pipes. Building owners also modified and carved up buildings to maximize the number of units (and their profits), which created dangerous conditions and serious building code violations. This practice is known as "milking," and it is one reason why properties in communities of color were often in such bad shape.[14]

Wealth Extraction

The government eventually passed laws to abolish redlining and end housing discrimination. But the real estate industry had another racist trick up its sleeve—"predatory inclusion," a practice that targeted African Americans for loans with additional fees and penalties, high interest, and adjustable rates.[15] While the legalese was different, little changed in housing conditions. The homes that realtors marketed and sold to African Americans were often dangerous and just as often squalid.[16] Yet this did not stop governments from systematically over-assessing properties in communities of color, extracting further wealth from minority homeowners and rendering them more vulnerable to tax delinquency and property liens—legal claims against property that city governments auction off to the highest-bidding investor.[17] In these ways, city governments enabled investment firms to profit from struggling homeowners like those who used to live on Loomis. Prior extraction of wealth and disinvestment coupled with deindustrialization concentrated poverty on the South and Southwest Sides of Chicago. Like in the South Side Chicago neighborhood that William Julius Wilson made famous,[18] jobs around Loomis Boulevard disappeared. In the 1970s, a public transit depot near Loomis closed, adding to the loss of stockyard and railroad jobs. Unemployment rose, and those that could left, taking their tax dollars to other neighborhoods and leaving behind them empty buildings and shuttered businesses.

Predatory inclusion ramped up again in the late twentieth and early twenty-first century, after regulatory changes to the way mortgages worked.[19] Historically, homeowners had taken out loans from—and then paid them back directly to—a lender. This meant that lenders were careful about who they lent to. In the new system, however, lenders were permitted to sell mortgages to investment banks.[20] Lenders stopped caring who could repay and offered (often aggressively) riskier loans. Investment banks preferred subprime loans because they have higher interest rates. As a result, lenders targeted low-income and minority populations, hedging bets that they would default on their loans and increase industry profit margins.[21] These risky loans—packaged as complex securities—were sold multiple times throughout the financial system. Home prices soared, and investment banks and financial institutions made huge profits, most of them from exploiting low-income people of color.

Foreclosures and Speculation

As is well known, this bubble burst. The lack of regulation or accountability for rating agencies, the increasing complexities of securitization, and the sheer quantity of subprime loans—originated to feed investors—spread the riskiness of subprime loans throughout the global financial system. When interest rates increased in the mid-2000s and the housing market began to stall, homeowners were left with unaffordable loans. A foreclosure crisis ensued.[22] The government bailed out banks and mortgage companies, protecting them from the ill effects of the crash. The same protection was not extended to homeowners like those on South Loomis, which led to the concentration of foreclosures and abandonment, and explains the oft-depicted derelict buildings, vacant homes, and empty lots in many low-income and minority neighborhoods.[23] "Poor people had nothing and they took all their shit," remarked a Southwest Side resident who lives near Loomis Boulevard.[24] There are thousands of foreclosed homes on the Southwest Side of Chicago.[25] And in 2016, between 10 and 25 percent of housing stock was empty on Loomis and the surrounding area.[26] The vast majority of neighborhoods with the highest recent rates of foreclosure are majority Black or Latinx.[27]

Foreclosures are bad news beyond the implications for their foreclosed owners because they exacerbate the deterioration of housing conditions. Banks can sell foreclosed properties at auction or opt out of assuming ownership. The latter often occurs in poor neighborhoods where banks want to avoid responsibility for tax payments and property maintenance.[28] Predicting sales will be slower and less lucrative in poor neighborhoods, banks allow buildings to become "zombie properties," where there is no clear owner. "Zombie properties" are properties on which a foreclosure claim has been filed but not resolved within three years. Because neither the lender nor the homeowner has control over the property, both lack clear incentive to invest in the property, pay taxes, or keep up with maintenance.[29] By leaving properties vacant and unmaintained, banks save money and avoid demolition costs. Zombie properties are disproportionately concentrated in lower-income minority communities that look like Loomis Boulevard.[30]

In addition to problems associated with physical dilapidation and increases in insurance premiums for neighboring homeowners, vacant buildings have distinct effects on their surrounding communities. Empty buildings and vacant lots are a moral and reputational force that local elites and outside investors can deploy for financial gain by using them as a focal point to justify new investment.[31] This problem is amplified because city governments tend to allow the owners of vacant properties a lot of time before they intervene.[32] Financial speculation turns vacancies into rent gaps, which are a prerequisite for economic investment and gentrification.[33] At financially opportune times, property acquisition companies and investment firms—like the New Jersey investment firm that now owns the two-flat on South Loomis—buy up high volumes of dilapidated buildings, including single-family homes, convert them into rentals, and turn them into real estate assets for private equity and other financial firms.[34] The deck is stacked, in other words, by and in favor of these financialized housing market actors.[35] These profiteers are part of a broader group called rentiers: people who profit from income derived from assets with a limited supply, such as housing.[36] In turn, profiteers have stacked the deck against homeowners on South Loomis, and thousands of others in Chicago.

Race and Real Estate

The stacked deck is also steeped in racism. While histories of slavery, racial capitalism, and exploitation created economic disparities between White

Americans and others, contemporary housing policies and practices sustain patterns of racial and economic inequality. For example, governments continue to over-assess homes in communities of color. This means that property owners of color pay higher taxes than Whites for comparable homes.[37] At the same time, the real estate industry consistently undervalues properties in communities of color, making it harder for property owners to build wealth and equity.[38] The real estate market also includes the people that articulate demand for housing—their individual behaviors play a role in stacking the deck as they buy into it and sort themselves racially. The demands of home buyers—for neighborhoods with good schools and reputations, for instance—contribute to the undervaluing of historically disinvested neighborhoods because they are less likely to have resourced schools or good reputations in comparison to White neighborhoods.[39] But realtors also make assumptions about where people want to live and steer White people to White neighborhoods, for example, further cementing segregation.[40]

The picture has been made bleaker still by banks that consistently deny conventional loans to homeowners in communities of color. Chase Bank recently loaned nearly nine times more in a single majority White neighborhood on Chicago's North Side than it did in all the city's majority Black neighborhoods combined.[41] Such obvious deck-stacking has severe consequences. Black and Latinx homeowners not only have lower home values; they are also more cost-burdened by their homes[42] and are much more likely to live in substandard housing. Their relative lack of wealth also results in less money for home repairs.[43] In many cities, unresolved building violations—like the issues in the house on South Loomis—result in property liens, further stacking the deck against owners of color.

Much of what South Loomis has experienced are national phenomena—from White flight, redlining, wealth extraction, and foreclosures to the pervasive over-assessing and undervaluing of communities of color. These policies and practices mean that the deck is stacked most overwhelmingly against residents in communities of color across the US. By contrast, banks and many Whites benefited from White flight, investment, positive appraisals, and increasing home values in the neighborhoods in which they live. Both sides of this history have created depressed housing markets on the South and West Sides of Chicago that result in the kind of disrepair, with pockets of speculative investment, found on Loomis Boulevard. History has not only created this material landscape; it has also created a situation in which this area has come to be known and recognized in terms of physical features of dilapidation, neglect, and abandonment. This is only one side of the coin, however, in Chicago's relational stacked deck. We visit another side next on our tour as we travel to Chicago's North Side.

Concentrated Wealth on the North Side

It is hard to miss the disparity if you go directly from Loomis Boulevard to Victoria Street. The trees in the yards on Victoria provide shade for the wicker chairs on the porches of houses, which are set back from the street in a leafy neighborhood by the lake on the North Side of Chicago. A couple—White and in their late fifties—have owned one house on Victoria Street for over twenty years. They paid off their mortgage early, which means they own their house outright and will never face mortgage foreclosure. The couple bought the house for a price equivalent to $86,000 in 2020 dollars. It is now valued at $790,000. The house is a few blocks away from homes that cost double that. To be sure, the house on Victoria Street is not a mansion. Compared to the residents on South Loomis, the owners are wealthy, but my bet is that they do not consider themselves rich. Their wealth is undeniable, but it is also subtle—more about stability and comfort than opulence. This is important because it makes their wealth less noticeable.

The house is impeccably maintained. The maintenance, plus a couple of remodeling projects through the years, have added thousands—probably tens of thousands—to the house's value. Homeowners with access to financial resources have a way of preserving and extending wealth through remodeling, building additions, keeping on top of maintenance, and other actions that add to property values. The house on Victoria is old, and there have been issues over the years that required permits and considerable repair work. But the couple could afford it—in part because of their salaries and in part because of some inheritance. There is no rotting siding or flaking paint.

The house on Victoria Street has not avoided building inspections, however. A few years ago, a neighbor called the City to complain about noise from a remodeling project. An inspector came out and found that the project was not permitted—the couple had not gone through official channels to get a permit for the work. As a result, the owners ended up in building court. As they recounted the story, it became clear that this was a headache for them, but did not cause an issue overall. "We went to . . . court and of course, we had to get the permit," they explained to me. "I mean that's the rules. . . . That was no big deal." The owners' nonchalance about the court case and associated costs reveals a lot about their access to resources. If you search for their property on the City's online building violation database, you can see that this one violation is listed as "addressed," suggesting that there are no remaining issues with the building.

> Submit plans prepared, signed, and sealed by a licensed architect or registered structural engineer for approval and obtain permit. ADDRESSED
> —Building inspector report for the house on Victoria Street, 2008

White Wealth

The couple's college-aged kids will inherit the house. But the intergenerational wealth that this hundred-year-old house provides is no stroke of luck. While federal policies subsidized White homeownership in the suburbs,[44] municipal policies also favored White neighborhoods and benefited White homeowners within cities. Propitious housing policy is one of the main reasons that White families now have nearly ten times more wealth than Black families. Much of the racial disparity in intergenerational wealth in Chicago can be explained by the thirty-plus-percentage point gap between Black and White homeownership.[45] Contemporary policies like the mortgage interest deduction program continue to subsidize and reward the most affluent property owners by allowing them to reap higher tax returns the more expensive their property is.[46] Wealth begets wealth, especially for homeowners.[47] The deck has been stacked in favor of wealthy homeowners and their children.

This house on Victoria Street shares a history with thousands of other single-family homes on the North Side of Chicago. Like many others, the neighborhood where this house stands was developed for wealthy Whites in the 1880s and has remained predominantly White to this day. Zoning decisions carved out White neighborhoods for residential use, while designating commerce and industry in and near communities of color.[48] Additional spot zoning decisions, annexation, and municipal incorporation also deterred integration, helping to shore up White spaces and White property values.[49] Even places on the North Side that were dilapidated and redlined in the mid-twentieth century recovered—due to conservation movements led by White middle-class "rehabbers," who sought to extend "already-established strongholds of the upper middle class."[50] Rehabbing on Chicago's North Side in the mid-twentieth century was not just about saving old, dilapidated buildings—it had just as much to do with avoiding poverty and growing populations of Black and Latinx residents.

Proximity to wealth—and distance from poverty—was also important on Victoria Street. So much so that, in the 1980s, neighborhood property owners persuaded the City to recognize the neighborhood as separate from an adjacent area that was distinctly lower-income and less White, thereby dividing the allocation of resources and changing the neighborhood's name. This "rebranding" allowed residents and real estate actors to market the neighborhood in a way that increased property values and concentrated wealth, which also led to better-resourced schools and other amenities.[51] These factors also make the neighborhood more desirable. The home-building industry is tuned into desirability and concentrates new construction in or near White areas. On top of new homes, uneven geographies of new construction mean that new drains, sewers, sidewalks, and amenities tend to be located in White neighborhoods.[52]

The concentration of wealth in many contemporary White neighborhoods also makes possible safe, comfortable, and even luxurious housing conditions: from lead abatement, modern HVAC systems, and manicured yards to pools, security systems, high fences, and gated communities. Some of these features—most often found among upper- or middle-class White households—are products of defensiveness and homeowners' racist perceptions of danger.[53] In other words, the condition and aesthetic of White neighborhoods are as much a product of urban policies and historical trends as those in communities of color. Yet the privilege that advantages White neighborhoods is obscured, and instead we are told that White neighborhoods are the default, desirable ideal, and conditions in communities of color need explaining. As Dianne Harris demonstrates, Whiteness and what we think of as "ordinary homes" grew up together. But architecture is not benign, especially when it is "so ordinary we hardly notice it."[54]

We know a good amount about the construction and co-option of aesthetic preferences of gentrifiers, and that some built environments are marketed as authentic, improvements, or as universally desirable when in fact they are not.[55] Yet the built environments of White homeowners—wealthy or not—have received less investigation from urban sociologists. White wealth accumulation is hidden, and we let houses in White neighborhoods be proffered as an ideal to all, despite their origins in discrimination, subsidies, and privilege. Housing conditions on the North Side are a product of decision after decision to prioritize White neighborhoods at the expense of communities of color. White neighborhoods are an integral piece of the stacked deck because local and national governments continue to prioritize their financial interests at the expense of other urban residents, whose neighborhoods and interests they neglect.

The stacked deck manifests in Chicago's geography. Most of the North Side looks different from the South Side and the West Side. The stacked deck creates and manifests as building conditions that are consistent within but widely variable across geographical areas of the city. Their shared histories of distinct treatment have created relatively discrete appearances in different neighborhoods. These material conditions become recognizable and socially significant—they can be read and interpreted. Yet, as the following chapters show, we do not always accurately recognize or interpret material conditions. One reason for this is that the reality is not always black and white, and contradictions and differences exist within neighborhoods, even though their material conditions are powerful symbols of wealth, poverty, and disparity.[56] There are pockets of wealth, development, and investment on the South Side, for example. One such pocket is the next destination on our tour of Chicago's housing landscape.

Pockets of Profit on the South Side

We go back south for the next stop on our tour. This time, though, we are directly south of Chicago's downtown, on the lakefront. Like so many others, this neighborhood experienced White flight, redlining, and disinvestment. But reinvestment, in part from a community bank, in the 1970s helped the neighborhood build and maintain a middle-class African American community and avoid institutional disinvestment.[57] A divide exists in this neighborhood, however, as Natalie Moore and Carlo Rotella describe, between the haves and the have-nots: between the proud owner-occupiers of bungalows, the lower-income tenants of apartment buildings (which house the highest number of housing voucher holders in Chicago), and residents in the strip of lakefront high rises.[58]

Luxury Living

It feels more like a hotel than a rental property. Views of the lake, twenty-four-hour fitness center, a dog run, a boutique lobby, and a party room with a fully equipped kitchen and bar, as well as doorman service. Lots of rehabbed rental buildings look like this. The uniform urban residential architecture has its origins in the 1980s and 1990s, during which time developers and city governments jumped on the opportunity to renovate old buildings and market them to affluent newcomers who valued the architectural styles and features of old buildings.[59] This helps to explain the homogeneity of building conditions in apartments constructed and targeted at young professionals,[60] and the neighborhoods—often close to cultural amenities and public transit—in which they are located. A building like this thus perhaps makes sense for this neighborhood, where the numbers of PhDs significantly exceed Chicago's averages;[61] developers of the lakefront high rise in South Shore likely had in mind upwardly mobile, middle-class, educated renters. But the number of neighborhood residents who did not complete high school also exceeds the city average.[62] And the building only recently became a luxury apartment building. Ten years ago, it looked quite different, as this building violation record attests:

1. exterior wall repair
2. north elevation immense structural wall damage window head cracking many pieces are temporally stabilized dangerous and hazardous
3. north elevation stone portion of facade spalling, damaged, structural cracks, pieces missing dangerous
4. northwest corner structural damage at brick wire meshed in place temperately dangerous
5. northwest corner parapet wall stone pieces have shifted, loose dangerous
6. all elevations severe washed out mortar, spalling brick, spalling concrete pieces loose dangerous
7. all elevations steel shelf angles rusting causing severe damage to surrounding brick and masonry
8. all elevations steel lintels rusting causing damage to surrounding masonry
9. west elevation severe facade damage window headers severely structural damage many temporary stabilized with wire mesh dangerous
10. west elevation old balconies temporary stabilized, makeshift enclosers many pieces loose, damaged, balconies originally open now enclosed and added as part of units. does not weight code loads
11. south elevation spalling concrete, damaged brick, areas wire meshed temporary dangerous
12. south elevation window heads damaged cracking dangerous

13. east elevation severe structural facade damage window headers cracking loose, some pieces are temporally stabilized dangerous
14. east elevation old balconies temporary stabilized, makeshift enclosers, many pieces loose, damaged, balconies originally open now enclosed and added as part of units. do not meet city weight code loads
15. canopies required at north elevation
16. penthouse lintels rusted, brick damaged, cracking, severe washed out mortar
17. many unit water infiltration causing damage to interior of units by facade deterioration units
18. plans and permits for all repairs
19. all elevations windows rotted / rusted

—Building inspector report for South Shore Drive high rise, 2012

The high rise clearly had issues—severe structural damage, dangerously cracked exterior walls, precarious balconies, leaks, and rotten and rusted windows. But despite these problems, there was profit to be made from this building. Let's begin, though, by taking a step back to understand how things got so bad.

The South Shore Drive high rise was built in the late 1920s and operated as rental apartments until it was converted to condos in the early 1970s.[63] Condos are different from rental buildings because people own individual apartment units. Owners pay homeowners' association fees every month to cover costs of things like building maintenance and water. And a condo association, comprising a group of condo owners, is responsible for distributing the money to cover monthly expenses and other costs that arise. By 2015, the South Shore Drive high rise was in a sorry state. A broken boiler and frozen pipes meant heat was not working, and residents were using their stoves for heat in the harsh Chicago winter. Ceilings were leaking, windows rotting, and masonry crumbling—especially problematic for a sixteen-story building. In court, an inspector testified that the building was a fire hazard because of issues with the gas pipes. But the condo association could not afford to make repairs. Half the units were empty, meaning the association was only collecting 50 percent of assessments. In addition, owners were suddenly responsible for paying a large fee due to a "special assessment" for necessary upkeep of the huge, almost hundred-year-old property. Condo owners who were able to sell were forced to settle for low prices. In 2014 one of these condos sold for $23,000. The owner had purchased it in 1979 for $43,000.

Condo Fraud and Deconversions

Condos have been making the news. During the housing bubble, fraudsters converted rental apartment buildings into owner-occupied condos. But they did so only on paper, and not in reality. Building owners, appraisers, lenders, and straw buyers[64] worked in cahoots to secure financing and advertise and sell condo units that did not exist or were already inhabited. Then they fled with the money.[65] Sometimes entire buildings were invented. In one case, the FBI indicted Chicago real estate investors and loan originators for their role in fraudulently obtaining mortgages totaling $8.5 million.[66] In other instances, management companies and other signatories of condo associations committed fraud by collecting—and then pocketing—assessments and fees, leaving condo associations without the funds for maintenance and emergency repairs.

In Illinois, as elsewhere, fraud prompted new legislation to mandate that condo buildings be "deconverted" to rental buildings to make buying and selling the buildings more feasible.[67] After the declaration of deconversion, the City files a separate motion asking for the property to be sold, oftentimes to an entity called a receiver. Receivership is a broad city initiative in which a building court judge appoints a third party to fix building code violations if the owner is not able or willing to do so. To ensure payment, the receiver records liens against the properties. Liens stipulate that the receiver can take possession of the property until the debt for repair work is paid. Thus, when owners do not pay, receivers foreclose, take possession, and sell buildings.[68] Though private receivership occurs, receivers in Chicago are usually public-private entities, funded by banks and subdivisions of the Treasury Department.[69] Receivership most often occurs in multi-unit buildings on the South and West Sides of the city, with some in the far North and Northwest.[70] While receivership ensures that repairs get made to buildings inspectors deem dangerous, it also enables profit by exploiting rent gaps in lucrative areas. In this way, receivership can amount to government-sponsored flipping.

Matching Profit Strategies to Markets

A building court judge appointed a receiver for the South Shore Drive high rise in 2015, after the building had been in building court—as a result of its long list of building code violations—for over twelve years. All other options were exhausted: the condo board was bankrupt, and no one had the money to address all the repairs that were necessary. Condo owners in a different neighborhood might have been able to bounce back from the housing crash, but this would have required more wealth in reserve than South Shore residents typically have, relative to their counterparts in majority White neighborhoods.[71] The receiver fixed up the building's exterior, removing imminent danger from falling bricks and balconies, and the $14 million rehab project was underway.

As soon as the initial work was completed, a transnational company snapped up the lakefront high rise. The company specializes in the acquisition, development, and management of rental buildings, and they own over 5,000 apartment units in the Midwest. While their profit margins are not publicly available data, this kind of landlordism is big business with an increasing presence in Chicago's rental market. At this writing, one out of every four homes sold in the Chicago area was going to investment buyers (i.e., owners who do not live in the property).[72] Whether they are buying up foreclosed single-family homes in bulk as portfolio or multi-unit buildings, these corporate landlords share a focus on maximizing returns for their investors.[73]

While replacing much of the plumbing and the boiler, the rehab project also converted the South Shore Drive high rise from forty-nine units to over one hundred rental units. After this work was done, a 428-square-foot studio unit rented for $1,249 per month and a 1,034-square-foot "penthouse" cost $3,500. Meanwhile, the median rent in Chicago was $1,077.[74] The average unit in the building is 850 square feet. It has become common practice to downsize unit square footage and include luxury amenities to rent properties at prices that are high for the area, with wide profit margins for investors and developers. The justification for the uptick in such small units is the provision of affordable housing, contemporary trends toward small households, and matching tastes of the young urban professionals who value location over unit size. But the high rents of many of these units—including those in the South Shore high rise—suggest that they are an efficient way to make money rather than a source of affordable housing.[75] It is no accident that units in newly rehabbed buildings across the city—and the nation—look similar. Size and style are products—as they always have been—of money-making strategies.

However, strategies for maximizing profit depend on local context.[76] In buildings like the lakefront high rise, property flippers and rehabbers try to make a quick profit by successfully predicting market appreciation in certain areas.[77] The luxury aesthetic and condition of the lakefront high rise is due in no small part to the fact that the surrounding area is coveted: it is by the lakefront, close to downtown and historic South Side neighborhoods, and near the site of the future Obama Presidential Center. In relatively strong housing markets like this one, landlords tend to upgrade properties: renovating, flipping, and raising rents. In weaker markets, landlords take aim at the low-rent sector of the market by intentionally undermaintaining properties while still increasing rents to exploit low-income populations with limited options. Landlords buy buildings but do little to improve them. Renting out substandard and dilapidated units can still turn high profits.[78] While some landlords cannot afford maintenance,[79] others—sometimes referred to as "milkers"—buy up cheap properties and intentionally spend as little as possible on rehabilitation and maintenance while maximizing rents.

Early work on slumlords, and the government's failure to prevent them, suggests that there is little profit to be made in slum housing, therefore owners are not economically capable of maintaining buildings.[80] In fact, scholars suggested that the widespread assumption about profit and willful negligence is the reason that slumlordism continues: the negative reputation prevents the government from assisting slumlords in making improvements. Recent studies, however, show that government policy encourages and enables huge profits from poorly maintained rental housing, a trend that is particularly prominent among professional and non-local landlords.[81] In strong *and* weak housing markets, landlords find ways to make profits for their investors, doing their best to ensure that the deck is stacked in their favor whatever the context. Landlords' actions vary across markets, but they share the capacity to produce physical conditions. As such, buildings in neighborhoods—or pockets of neighborhoods—with strong housing markets tend to look one way, and those in weak markets look another.

Subsidized housing has some particular aesthetics too. The South Shore Drive high rise stands among apartment buildings that are home to the highest number of housing voucher holders in the city. Housing vouchers are one piece of federal housing policy. The demolition of public housing and other policy decisions have created a market in which private entities can profit—often through tax incentives or the guarantee of rent dollars from the housing choice voucher program (commonly known as Section 8)—from the provision of subsidized affordable housing units.[82] Landlords that rent to voucher holders also play their part in creating specific housing conditions. As Eva Rosen describes, landlords sometimes add additional layers of drywall to cover lead paint, decreasing square footage, because they know housing inspectors look for lead paint in units rented to voucher holders.[83] During renovations, housing choice voucher landlords remove second bathrooms, windows, and basement access so that tenants have fewer toilets to clog, windows to break, and space to inhabit. Landlords' assumptions about voucher holders' behavior shapes the material conditions of their housing. As such, though voucher holders rent in the private rental market, the units available to them may not look like an average rental unit, and certainly do not resemble the upmarket rentals in the South Shore Drive high rise. Here, as in many other places, disparity is as evident within the neighborhood as in comparison with others.

Displacement on the Northwest Side

Next on the tour, we travel to Wellington Street, in a neighborhood on the city's Northwest Side. Historically, this has been a first destination for immigrants and refugees, who settle in the neighborhood before moving to the suburbs.[84] Over time, the neighborhood's housing stock of apartment buildings and two- and three-flats constructed in the early 1900s has been home to large populations of Jews, Koreans and Filipinos, and Mexicans. Neighborhood organizations believe the housing stock is one reason the neighborhood was so popular with working-class immigrants: the three- or four-bedroom flats include a living room that can be converted into an extra bedroom, making the neighborhood ideal for cohabitation or doubling up to save on rent.[85] In the 1970s these aging buildings caught the attention of city planners, who designated parts of the neighborhood as "blighted" and prime candidates for urban renewal. Residents and community organizations stopped this from happening, and the area remained largely untouched by city planners or developers. Instead, the government and local development corporations joined forces to fix up dilapidated buildings and offer low-interest loans.[86]

Now the Northwest Side neighborhood is at the forefront of political conflict over gentrification in the city. It is close to public transit and abuts other popular neighborhoods that have become increasingly expensive. While the term "gentrification" is used by activists, residents, community organizations, and the media, its definition has been notoriously hard to pin down. This is in no small part due to competing images of gentrification. Some see gentrification as a nearly unstoppable force that entails displacement and drastically determines residential patterns in cities. Others are more hesitant, and see gradation in the neighborhood changes included under the umbrella label of gentrification.[87] Unstoppable or not, things have changed in the neighborhood. Condos have become more common. Rental buildings are increasingly owned by limited-liability corporations (LLCs), which means owners are protected from personal financial risk and thus have less incentive to provide decent housing or to follow the letter of the law in managing property. LLCs became available in the 1990s, and their presence in the Northwest Side neighborhood has coincided with intimidation tactics to displace tenants, and increased rents.[88]

Tools of Displacement

Targeted inspections of buildings that have been identified as problematic—usually for alleged drug and gang activity—are another tool of displacement.[89] These kinds of inspections happen in cities across the country. In Chicago, they are called Strategic Task Force (STF) inspections.[90] Spurred by a request from the police department, a task force enters (forcibly if necessary) a property linked to criminal activity. The task force—comprising a team of building and fire inspectors, chaperoned by the police—combs through the building under instructions to cite the owner for every possible violation of building and fire codes. There were 5,705 STF inspections between 2006 and 2015. The aim of these inspections is to cite property owners for as many violations as possible, send the building to court, and pressure the owner to evict tenants deemed problematic. Crackdowns on so-called problem buildings are part of community policing programs in cities nationwide, which, due to race- and class-based biases about disorderly behavior, tend to disproportionately penalize minority residents and regulate minority neighborhoods by enforcing behavior that is acceptable to White middle-class moral standards and removing "problematic" minority residents.[91] Indeed, as the map in appendix C shows, STF inspections are concentrated on the South and West Sides of Chicago—areas of the city in which communities of color are most concentrated.

> Front exterior entry door—deadbolt lock missing [building unsecured]. Front interior stairway with unsanitary carpeting, graffiti on walls, and trash on stairs. Front doors do not lock, and homeless people sleeping in hallway.
> —Building inspector report for Wellington Street four-flat, 2019

The above building code citation is just one of the twenty-three different entries a building inspector reported during an STF inspection of a four-unit building on Wellington Street, in this Northwest Side neighborhood. Other violations in this building include broken windows, missing smoke detectors, kicked-in doors, faulty wiring, and insufficient plumbing. Community members and the police had implicated tenants in this Wellington Street four-flat in a series of shootings and drug dealing on the block.[92] The ensuing STF inspection prompted a court case that forced the property owner to either evict the "problematic" tenants and fix up the building or sell it. STF-induced evictions are just one form of displacement, however.

A few blocks down Wellington Street sits a red brick six-flat with a host of issues. Exposed wires in the hallways and basement. No smoke detectors. A porch that had partially fallen. Mouse droppings. The tenants had called the City to report these issues. And it was their right to do so—these are violations of the city's building code, and legislation requires Chicago landlords to make these repairs and to provide smoke alarms.[93] But their building owner had done nothing. The tenants were hopeful when this inattentive owner sold the property. But the new owner also failed to deal with the issues. Rather than addressing the tenants' concerns, the new owner gave the tenants a thirty-day notice to move out. The tenants suspected that their eviction was retaliation for their complaints, which is an illegal, but common, practice.[94] With the help of a community organization, the tenants organized a rent strike. Again, the tenants were within their rights; tenants can legally withhold rent if repairs are not done.[95] Some safeguards exist for tenants in Chicago, and a handful of organizations offer tenants advice and legal assistance.[96] Yet the lack of rent control, horror stories about subpar housing conditions, and the proliferation of evictions in the face of increasingly profit-driven corporatization of the rental market mean that the deck is stacked against tenants despite extant protections.[97]

Condos that replace rental units on Wellington Street and the surrounding neighborhood lack the luxury of the lakefront high rise we visited on the South Shore, but they boast many more amenities and features than their rental predecessors: in-unit washers, dryers, and dishwashers; colored tile backsplash in kitchens; dark stained hardwood floors; and quartz countertops. Their exteriors are unchanged, aside from fixed up masonry and porches. But these are old buildings, whose age and historic character likely appeal to the upwardly mobile urban professionals moving into the

Northwest Side, just as historic architecture does to middle-class residents in Chicago and other US cities.[98] Aesthetic preferences and architectural choices made by developers structure the look and feel of middle-class housing in contemporary cities. And this has implications for affordable housing, stacking the deck against lower-income renters. Small properties like these have historically been a source of affordable rental housing. Now they are the only housing type in decline in Chicago as, in tight markets like this one, building owners and developers gut the interiors to build condos.[99]

The tenants from Wellington Street were some of the 200—mostly Latinx—families displaced from their homes in one three-square-mile area on the Northwest Side of Chicago.[100] Evictions happen for a variety of reasons—because landlords want to rehab buildings and raise rents and because tenants miss rent payments. But, across the board, evictions reproduce poverty by rendering previously evicted tenants more likely to live in substandard housing and making it harder for them to access housing because landlords screen for previous evictions.[101] Mirroring nationwide patterns, renters in minority neighborhoods are far more likely to face eviction than in Whiter areas of Chicago.[102]

Demise of Affordable Housing

Affordable housing is not just an issue for people with low incomes, and it does not refer only to government-subsidized housing. Housing is affordable if it costs someone less than 30 percent of their income. Affordable housing is increasingly hard to come by for middle-class residents in contemporary cities. Their limited choices also further constrain the choices of their lower-income counterparts. While some people argue that building more market-rate housing will naturally lead to a filtering process whereby some housing will be available at affordable rates, this does not work in tight markets in part because of middle-class preferences for older housing. In fact, areas like the Northwest Side are more prone to "reverse filtering," in which middle-class renters crowd out affordable housing, resulting in fewer affordable units for low-income renters.[103] Adding to this are unregulated rent increases and stagnant wages, which have created a dearth of affordable housing in many US cities.

Although housing has become increasingly unaffordable across the board, renters of color facing discrimination and citizens with criminal records have the fewest options, and the fewest resources to avoid exploitation in the rental market, and are at most risk from housing precarity. As in many other US metro areas, Black and Latinx residents in Chicago are more likely than Whites to a) be renters; b) live in substandard conditions; and c) be cost-burdened by their rental units.[104] Black residents are more likely to have criminal records—which landlords can legally use to deny rental applications. As such, the repercussions of increased rents are more likely to affect Black and Latinx renters, who have on average the least financial resources and whose housing access is more constrained than other groups due to the aforementioned issues. Black and Latinx renters are most likely to move into buildings with substandard conditions that receive violations in the first place,[105] even if landlords follow the rules by communicating information about code violations before tenants sign a lease.[106] While some of the tenants evicted from the Wellington Street area were priced out of the Northwest Side neighborhood and moved away from the place they had called home for over a decade, others reported that they moved into nearby buildings that were in worse condition.[107]

Tenants in the low-income rental market have fewer options, so there is less impetus for landlords to maintain buildings to keep tenants happy. The deck is most stacked against low-income minority renters and, oftentimes, in favor of those who house them. The deck is stacked heavily in favor of rentiers: those who derive income from the ownership, possession, or control of property under conditions of limited or no competition.[108] The way the deck is stacked creates distinct building conditions in different housing markets, and in different parts of the city. These are recognizable and enable categorizations of the city, whereby we can drive past a building that looks a certain way and glean something about who lives there.

Conclusion

The tour of Chicago's stacked deck has explained why, on average, buildings are in worse shape on the city's South and West Sides. This is especially the case for properties that house low-income renters or homeowners who were targeted for risky mortgages. By contrast, more money is spent on housing and its upkeep on the North Side. The stacked deck creates clear groups of building conditions in different parts of the city. Some aspects of this tour are particular to Chicago, and others depict more general urban processes. Rather than pick apart what is specific and what is general, the tour of the stacked deck reveals how policies and practices are reified in built form.

The stacked deck creates a legible landscape of disparate housing conditions. We see dilapidation or luxury, and we tend to classify places and people in these terms too. Buildings—and their physical condition—serve as signals that guide our evaluations of people and places. Relatively discrete appearances of different neighborhoods become recognizable, and therefore socially significant; they can be read and used to infer a great deal about who lives there, who belongs, who should avoid a neighborhood, and how the deck is stacked. Building inspectors work within this stacked deck. They drive down streets that span the city, take shortcuts, make wrong turns, get lost, and have to retrace their steps. They get called to buildings in total disrepair as well as to new luxury properties. They see it all. And they use the stacked deck as a categorizing schema to inform their opinions and decisions about who deserves leniency or punishment.

In the next chapter, we find out more about how this city of stacked decks motivates building inspectors to make surprising decisions. We learn about the importance of their residential backgrounds, their informal training, their awareness of and involvement in headline-making scandals, their on-the-job discretion, and their encounters with the gamut of wealth and poverty. But their decisions about building code violations and housing conditions present us with a puzzle. Their assessment of people and property only partially aligns with the reality of housing inequality.

CHAPTER TWO

Building Inspections

We could see for miles from the fourth-floor window in the Buildings Department office. Red brick low-rise apartment buildings dating to the early twentieth century made up most of the view, but there were newer looking buildings here and there too. "I could find violations anywhere," inspector Antonio told me as he looked out onto this scene, "even in brand new buildings." Antonio, a tall Latinx man with the beginnings of a beard, is a building inspector in his forties.[1] His statement was a common refrain among Chicago's inspectors: all buildings have violations. But inspectors do not assess all buildings in the same way. They do not record or follow up on all the violations they see. Nor do they always overlook the same issues in every building or neighborhood.

To an outside observer, it may seem like there is no rhyme or reason to inspectors' decisions. But there is. There is both rhyme (i.e., a pattern) and reason (i.e., motivation) to what inspectors do. In this chapter, we find out that the stacked deck drives inspectors to make their decisions. Hearing more from inspectors like Antonio, we will begin to understand how class backgrounds, professional recourses and clout, and experiences motivate inspectors' particular view of the stacked deck. We learn what inspectors do during inspections, how they decide what to do when their inspections turn up violations, and how they continue to exert discretion after their inspections are over. We discover how the building code enables, guides, and constrains inspectors' discretion and find out about the courses of action—from citations to court cases—that are available to inspectors during and after inspections. The chapter ends with a discussion of how the stacked deck shapes frontline work in other contexts. But before we dive into the ins and outs of building code enforcement and frontline work, we need to get to the bottom of a puzzle.

The Puzzling Geography of Building Violations

As the tour of the stacked deck in the previous chapter reveals, housing conditions vary dramatically across the city. We can see this pattern reflected in figure 4—a map of 311 service requests that residents made to express concerns over building conditions. Chicagoans made 260,474 of these service requests about buildings between 2006 and 2015.[2] And the map shows the concentration of calls per census tract for this ten-year period.[3] Concerns range from structural damage at a neighboring property and problems with rodents, leaks, and mold, to tenants calling about landlords who will not make repairs or who fail to provide heat in the winter. The map depicts the rate of requests per housing unit in an area. This means that the darker areas on the map are not a result of there being more housing units in those places (and thus more issues or more people to call); rather, darker areas mean that the rate of requests per housing unit is higher.

There are more 311 requests about building conditions per housing unit in the south and west of the city. This mirrors what we learned in chapter 1: these areas have long histories of disinvestment and more poverty than do the north and center of the city. Thus, it makes sense that buildings in these places are more likely to be dilapidated and poorly maintained. But, as the map shows, requests come in from across the city—from wealthy neighborhoods too. While property owners tend to have more money in wealthier neighborhoods, buildings still age and fall into disrepair, owners still do work without permits, and landlords still neglect maintenance.

Building inspectors follow up on these 311 requests and decide if buildings have code violations. Yet, a map of building violations—the issues that inspectors record—looks quite different from the map of 311 requests. The map in figure 5 shows building violations per 311 request for each census tract in Chicago.[4] In other words, figure 5 shows how many of the requests mapped in figure 4 translate into violations, and where. Put another way, it shows whether 311 requests amount to building violations. Census tracts in this map are darker not because there are more requests, but because there are more recorded violations. Thus, the map captures inspectors' assessments about building conditions and not just the geographical distribution of 311 calls.

Despite some relative consistency, the maps tell us that, usually, more requests do not equal more violations. Indeed, some areas of the city look quite different in the two maps. While there are some general commonalities

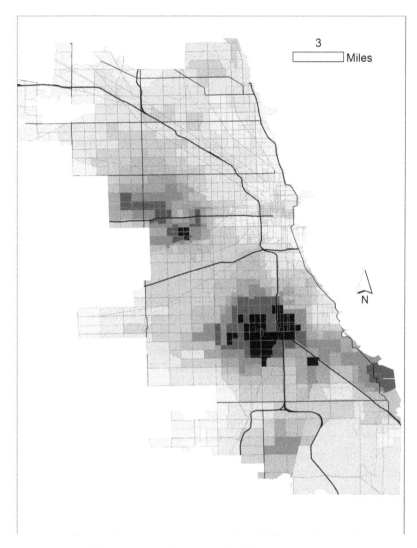

FIGURE 4. 311 building requests per housing unit in Chicago, 2006–2015. Darkest areas are well above mean, lightest areas are well below mean. Data are shown by census tract. Map by Richard Campanella.

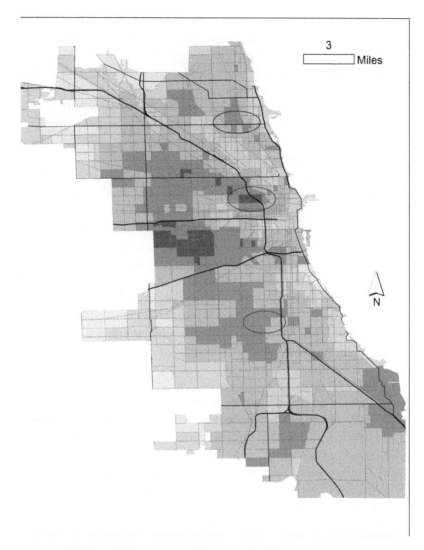

FIGURE 5. Building violations per 311 building request in Chicago, 2006–2015. Darkest areas are well above mean, lightest areas are well below mean. Data are shown by census tract. Map by Richard Campanella.

in pockets of 311s and violations on the South and West Sides, for example, inspectors seem to issue more building violations on the North and West Sides than we would expect from the number of 311s in these areas. This means that people are calling 311, but often their calls do not amount to recorded building violations. There are high volumes of recorded building violations in parts of the city not only because the buildings warrant code violations, but also because of the discretion building inspectors deploy. So, if we want to understand the differences between these two maps, we need to find out how inspectors make decisions.

I discovered that inspectors use the stacked deck as a tool to guide their decisions. In other words, they use their perceptions of the relative inequalities they see throughout the city. In chapter 1, we surveyed what we know from historians, sociologists, and others about the stacked deck. But people's perceptions of the stacked deck are not as complete as these accounts. And it is *perceptions* of stacked decks that motivate inspectors' decisions. When they assess a building for violations, they consider the position of the building or its owners within the stacked deck. And they have clear ideas about who wins and loses within the stacked deck. For inspectors, the deck is stacked by and for urban profiteers. It is stacked against many of the people profiteers exploit. Therefore, when they inspect rental properties, inspectors rail against big professional landlords who fail to keep up their properties, especially if they operate luxury buildings or charge high rents. By contrast, they show leniency to small-time landlords and family buildings because they believe these property owners are hardworking but struggle to get by within the rigged deck. But they do not empathize with most tenants, nor can they do much to help them. Inspectors try to protect low-income homeowners in communities of color, especially in the wake of the foreclosure crisis. Yet they combine structural and cultural explanations for racial disparities in housing conditions, and they have White blind spots, meaning that they overlook how Whites have benefited from the stacked deck. Similarly, they scorn government intervention that steers investment to neighborhood newcomers and stacks the deck against old-timers; however, in some changing neighborhoods inspectors protect newcomers as much as old-timers because they see gradations of investment in different contexts. Each of their decisions is informed by who they believe profits and suffers within the stacked deck.

We can see some of these patterns on the map in figure 5. Inspectors record fewer violations than average in some areas of the South and

BUILDING INSPECTIONS 53

Southwest Sides, even though these areas have more 311 calls than average. For example, there is a cluster of tracts with high rates of building violations on the near Southwest Side (circled on the map in figure 5). As we learned on the tour of the stacked deck in chapter 1, this area has a long history of wealth extraction and poverty, culminating in high rates of predatory lending and foreclosures. Residents in this area have borne the brunt of systematic racist disinvestment and exploitation; the deck has been stacked against them. Inspectors use their discretion to show compassion to people in this situation. Thus, when possible, they make stabs at justice by not recording violations unless they are dangerous. This explains the relatively low number of recorded building violations, despite the high number of 311 calls. And while it might seem like inspectors are making decisions that neglect legitimate requests about building conditions, understanding where their motivations come from reveals that their actions are actually attempts to dole out their version of justice.

Now let's look at an area on the city's North Side, also circled in figure 5. This area (like some others nearby) has a higher-than-average number of building violations, but a lower-than-average number of 311 requests. In other words, inspectors seem to record more violations than the volume of 311 requests would call for. What's notable about this is that inspectors are doing this on the North Side, which is majority White and wealthier than other parts of the city, and where landlords charge higher rents. Knowing what we do about inspectors being motivated by the stacked deck, this pattern is an indication that inspectors dole out as many violations as possible in areas of wealth and in high-rent buildings, in an effort to bring profiteers down a peg or two. The high number of building violations signifies inspectors' stabs at justice.

There is a third area circled in figure 5—on the near Northwest Side, straddling an expressway. The number of 311 calls there is well below the mean, but the rate of violations is well above average. This is part of an area—close to downtown, public transit, and the expressway—that has seen a lot of building renovations and upscaling.[5] Rents are higher in these newly renovated buildings, which used to be a source of more affordable housing in a central location. Knowing that inspectors disdain these kinds of developments and the profiteers behind them gives us a clue as to how we should interpret the high number of violations in this part of the city. Inspectors are using their discretion to punish property owners in this area—who benefit from the stacked deck, and in fact further stack it in their own favor—by citing them for as many code violations as possible.

The high rate of violations makes sense once we understand inspectors' motivations.

These are just a few examples of how inspectors exercise discretion, and the maps show them at the level of census tracts, rather than individual buildings. We'll read more about how very local contexts matter to inspectors in later chapters; it is important to know for now that inspectors do not just turn up in response to a 311 request and automatically cite the building for violations. If they did, we would see more consistency between the two maps. Instead, inspectors make decisions. Sometimes they do not record as many violations as we would expect from the number of 311 requests that come in. Other times they record more. Another important thing to know is that inspectors' decisions are clear verdicts on who, what, and where they think deserves leniency or punishment. Inspectors see a stacked deck that aligns partially, but not fully, with the patterns revealed on our tour of Chicago's housing landscape. They try to protect low-income homeowners in communities of color because the deck is stacked against them, but do not extend this protection to tenants. They come down hard on people who profit from poverty, but do not recognize the extent to which White residents have benefited from the stacked deck.

But where does their perception of the stacked deck come from? Why does it only partially align with how we know the deck is stacked? Why is there such consensus among inspectors about how the deck is stacked? The following section answers these questions by outlining three overlapping factors that shape inspectors' perception of the stacked deck: their social location, the recourses and clout they have on the job, and their interactions with the stacked deck itself. Taken together, these factors explain the way Chicago building inspectors interpret the stacked deck, make decisions, and shape the very city that they interpret.

Where Perceptions Come From

Social Locations

Like all frontline workers, inspectors are what Celeste Watkins-Hayes calls "situated bureaucrats,"[6] meaning their decisions and behavior are products of systemic social locations such as their class, race, and gender. About 80 percent of Chicago's building inspectors are White men. Of the remaining 20 percent, half are Latinx and half are Black. Women make up only a small handful of inspectors.[7] Thus, most inspectors are White men. The similarities do not end there. Inspectors also share a common class position—they are from working-class families and neighborhoods. Most did not go to college,[8] began their careers in construction, and became inspectors later in life. Like working-class men in other settings, inspectors value hard work, straightforwardness, and sincerity.[9] And they dislike people who are phony, cheat the system, and take advantage of others, as well as those who think they can pull the wool over your eyes and who interfere with "honest" people trying to "make it."[10] They enjoy David and Goliath–type stories where an underdog goes up against the evil giant.[11] These stances were pervasive among the White, Black, and Latinx inspectors with whom I spoke.[12]

Inspectors seemed unanimously proud of their jobs. They regard their careers in city government—complete with pensions, benefits, a steady salary, and freedom from physical labor—as "good jobs," steps up, and positions that they—as working-class men—feel lucky to have.[13] One inspector, for example, told me he "jumped at the chance to have a City job" when he heard the Buildings Department was hiring. It is also noteworthy that many inspectors were members of their regional carpenters' union, which may contribute to their sense of struggle and solidarity. Inspectors carry their support for the underdog and their respect for hard work and responsibility into their daily work lives. They seek to protect struggling homeowners and small-time landlords. Their disdain for cheating and exploitation causes them to punish unfettered profiteering and negligence.[14] In short, their stances translate into people and buildings, and their social locations help to explain the way inspectors perceive the stacked deck and categorize the city.

Clout and Recourses

Inspectors have clout. They enforce municipal regulations, which means that—unlike an average citizen—their actions are backed by the political and legal framework of municipal code enforcement. Their decisions and actions are impactful. They have the power to cause serious headaches for building owners, and often get the final word on what counts as decent or safe housing. But their clout is restricted by the specific set of recourses—ways of dealing with situations—available to them as inspectors. They can overlook issues, cite property owners, send cases to court, or recommend demolitions. They cannot, however, cite tenants for code violations, or offer property owners resources to make much-needed repairs. In fact, recourses within code enforcement are directed solely at property owners and are more punitive than reparative.

But selecting their recourse and wielding their clout first necessitates that inspectors make decisions about what to do when their inspections turn up building violations—and that means inspectors must make sense of complicated situations. This is a trait common to all frontline work. When welfare workers decide who deserves leniency and who deserves punishment, for example, they categorize clients using shared cultural stereotypes and assumptions.[15] Police officers also use shared moral assessments about who is good and who is bad,[16] and ambulance crews decide whether patients are legit or bullshit.[17] Their work necessitates that they use shorthand categories. For inspectors, classifications about who the deck is stacked for and against become salient organizing categorizations in the city. Inspectors slot people and places into these categories.

Frontline workers' assessments of who is deserving stem from institutionalized practices—"common ways of doing things that, while not required by any specific official policy, are supported and legitimated by rules, training, and law, and that spread widely to become a commonly accepted activity."[18] Individual frontline workers are able to exercise discretion, but they tend to make institutionalized and routinized decisions. Another way of saying this is that, although they make individual decisions, they regularly make the same decisions as their colleagues. This is true of building inspectors in Chicago: inspectors have clout and a set of recourses, but how they use them is institutionalized and patterned by informal training and the professional culture in which they work.

New inspectors typically train with more experienced colleagues and pick up on widely accepted patterns of discretion. Al, a White inspector, told me that "new guys" have "a habit of citing everything." Nick—who is also White and joined the Buildings Department around the same time as Al—concurred: "[new guys] want to try out every code in the book," he told me, "Like writing someone up for not having permits thirty years late"; but "old inspectors aren't going to harass people." Al illuminated the difference between old and new inspectors: "Down the road," he explained, "inspectors make decisions about whether they're going to write something or not write something." Yet there is no formal training involved, as Bill, a White supervisor in his sixties, explained:

> There's nothing in writing... it's just more from experience and it's just listening to other inspectors talk about what they've experienced during the day and somebody might say "well I came across an apartment, there was a mother with five children and it didn't have any heat" or "it was a mother and five kids living

in a vacant building," that kind of thing, and that inspector might have said "well, I called [the supervisor] right away and we had them ... the Department of Family and Support Services came out and relocated them ..." and somebody that could be listening to that story could think "oh, ok, that's what I'm supposed to do in that case." Because nobody ever told them.

As they gain experience, and receive nudges from their colleagues, inspectors begin to understand what is expected, and how to use the recourses available to them. During an interview in the office, Steve, a White inspector in his sixties, told me, "If you work in a lot of distressed neighborhoods, what is defined as bad can mean a lot of things ... the layman could go out to some place and think it's horrific, whereas I might think, well, it's not bad." After deriding a new inspector for citing a homeowner for a missing fence post, one inspector went on to tell me that his years on the job allowed him to distinguish between what matters and what doesn't. "With my experience," Paul, who is White and in his late fifties, told me, "Even when something appears dangerous ... I'm able to look at violations and see which things are important [and] which things are not important." Thus, we can begin to understand that, when an inspector cites a property owner for minor issues, this is an intentional and strategic decision that, as we will find out, is a form of punishment. The recourses inspectors choose—from those at their disposal—are shaped by their informal training.

Another factor that influences inspectors' day-to-day work is a concern with recent headlines about racism on the part of other frontline workers—the police in particular.[19] It is plausible that inspectors strive for political correctness, or elide discussions of race altogether, to avoid negative attention.[20] Consider inspector Bill's comments on new hires:

> New hires have to learn stuff about working for the City, like to be sensitive to race relations, equality, LGBT community. Some new hires ... are anxious about going to bad neighborhoods. Some guys come from environments where it's normal to use crude language. But working for the City means you do not have the authority to either discriminate or support your point of view, no matter what you think about certain issues.

Bill notes that new hires are taught to be "sensitive to race relations." I took this to mean new hires are told to avoid making decisions that could be construed as racially biased—and that this sets in motion an institutionalized pattern in discretionary decisions. Just as frontline workers can learn discriminatory practices that penalize people of color it can also become second nature to go easy in communities of color as an institutionalized practice to avoid negative attention.

There is evidence that informal training works. Supervisors tend to trust longtime inspectors and provide little oversight, rarely questioning reports. Inspector Nick estimated that his supervisor queried "one or two [reports] in every 1,000 cases." Likewise, the Buildings Department employs a rotation system, which means that inspectors' direct supervisors have all worked as inspectors, and some inspectors have been supervisors.[21] Inspectors thus make similar decisions—they pretty uniformly apply the same logic when deciding the extent of code violations that they will issue or aggressively pursue. The reason for this is, in part, to fall into line with their colleagues and the institutionalized practices of code enforcement in Chicago.

Interactions with the Stacked Deck

Inspectors' interactions and experiences with the stacked deck also shape how they perceive the city and the buildings they assess. Inspector Danny tried to articulate how experiences have changed his assessments over his career:

> I remember when I really got a leg up in figuring out how to write buildings up. I was like a bat out of hell in these neighborhoods because I could see so many problems. I was essentially told that I was really hurting people, and maybe I was. I don't think I was any less empathetic than I am now, although I've experienced a lot more.

Experience on the job and in the city changes inspectors' perceptions. But it also matters that they have experiences with material things. Inspectors spend their days looking at buildings, which signify socioeconomic status and resources as well as disparities in wealth and neighborhood context.[22] As Fran Tonkiss argues, urban inequality "gives itself away in space."[23] Eddie, a White inspector who just turned fifty, mentioned a revelation when a 311 call first brought him to a street of grand and ornate nineteenth-century red brick homes in a majority African American neighborhood. "I was surprised," he told me, "Because of how beautiful some of the buildings were. It was like punching a hole in a big lie. Not all

Black people live in ghettos." Inspectors' work entails assessing material conditions, and this can challenge their assumptions about race and the city.[24]

Inspectors' experience within the stacked deck generally predates when they began their jobs at the Buildings Department. Most White inspectors grew up in Chicago's Bungalow Belt—the series of detached single-family homes on the city's far South, West, and Northwest Sides. Although demographics are changing, these far North, South, and West Side neighborhoods have historically been home to many city workers who, despite any desires to move to the suburbs, live within the city limits because of residency requirements for city employees. The high rates of homeownership in these neighborhoods also mean that inspectors identify and empathize with property owners. And while they may not have grown up around landlords, they valorize the combination of hard work and small returns that they assume to be characteristic of the owners of small rental properties. They do not identify with renters in the same way, which helps to explain the lack of concern they articulate for tenants during inspections. The bungalows that characterize many of the neighborhoods in which inspectors grew up have also long represented a tangible expression of hardworking families and strong, reliable values,[25] which align with inspectors' working-class sensibilities and compassion for hard work and modesty.

These are the kind of residential environments that US society has been sold as the desirable—and achievable—norm.[26] Hidden from view is that this landscape is a product of a racially unequal state-sponsored housing market that encourages homeownership and wealth for Whites. As such, their history of growing up in this environment explains, to a degree, why inspectors do not realize that White residents—including people like their parents—benefited from the stacked deck. In short, inspectors' residential backgrounds help to explain what I call White blind spots—their lack of attention to Whiteness and White privilege. Black and Latinx inspectors, for the most part, also grew up on the South and West Sides of the city, but in majority Black and Latinx neighborhoods. While this means that they may be less likely to have White blind spots, it is also plausible that the ethos among inspectors—that of White working-class men—is potent enough to become the modus operandi for inspectors of color too.[27]

However, things have changed since inspectors' childhoods in the 1950s and 1960s. Whites left most of the parts of the city inspectors once called home, and many neighborhoods on the South, West, and Northwest Sides South are now predominantly Black and Latinx. What's more, legacies of housing discrimination have distributed material conditions unevenly, meaning homes owned by Black and Latinx Chicagoans are smaller and less expensive compared to those of Whites.[28] The distribution of people within the kinds of buildings they exalt—such as modest bungalows in the neighborhoods where they grew up—means inspectors are prone to extend benevolence in many contemporary communities of color.[29]

Inspectors' experiences in the city intersect with their social location and their recourses and professional clout to create a shared understanding of the stacked deck.[30] They use the stacked deck to make distinctions and draw categories. They are motivated to assist those they believe the deck is stacked against. Most often, this means struggling homeowners and small-time landlords. And, whether it is because their social location prompts empathy for marginalized populations, because their work environment makes them fearful of displaying racial bias, or because their interactions with racially disparate housing conditions provoke concern, inspectors are compassionate in communities of color. By contrast, they seek to penalize people who they believe benefit from an unevenly stacked deck, such as professional landlords and other urban profiteers; yet they overlook the extent to which Whites have benefited from the stacked deck. But how do they show leniency to some people and penalize others? What are the actions they take to protect struggling homeowners and punish big landlords? In the next section, we learn much more about what inspectors do.

Building Code Enforcement in Chicago

Building inspectors work for the Buildings Department, which is one arm of the Chicago city government. Most cities divide government work into different departments, and mayors appoint a commissioner and establish a budget for each department.[31] In Chicago, departments include the Department of Finance, the Police and Fire Departments, the Department of Housing, and the Department of Cultural Affairs and Special Events. Some departments, including the Chicago Buildings Department, generate revenue for the city through fines.[32] Chicago's Buildings Department is responsible for "enhance[ing] safety and quality of life for Chicago's residents and visitors through permitting, inspections, trade licensing, and code enforcement."³³ Code enforcement in Chicago entails a fleet of building inspectors, and a legal and bureaucratic framework that enables the City to penalize property owners for violations of the municipal building code.

Unkempt yards overgrown with weeds, rats, broken windows, peeling paint, crumbling walls, and rotting wood—these are the kinds of issues that are listed in the building code and that occupy the minds of building inspectors, and the pages of this book. Regulations about buildings are part of a broader category of land use laws that originated in the early twentieth century. According to Mariana Valverde, land use regulations emerged from the "fortuitous conjunction" of several unrelated urban and political trends, including Progressive campaigns against overcrowding and poor living and working conditions.³⁴ In many ways, the history of municipal regulation is also the history of US cities.

The origins of building regulations in Chicago date to the 1830s, when the city was rapidly developing with little in the way of precaution. Most houses were made from wood, which prompted concerns over fire safety. For this reason, early building regulations focused on restricting hazardous building materials and regulating design, operation, and maintenance of chimneys and stoves.³⁵ Time and again these regulations proved inadequate; Chicago was beset with fires. Partially in response to pressure from insurance companies (who were sick of paying out), the city council created a Buildings Department in 1875, complete with a team of building inspectors.³⁶ Campaigns around the turn of the twentieth century to reform housing conditions in tenements also inspired building ordinances concerned with sanitation, light, and ventilation. Soon the Chicago Buildings Department was inspecting the living conditions of the city's apartment buildings, in addition to enforcing fire codes.

As in other municipalities, Chicago's building code grew in response to events in the city. A series of tragedies over one hundred-plus years caused the City to adopt specific ordinances. After Chicago's Great Fire in 1871, the City got stricter about building materials and mandating open space between adjacent buildings. A fire in the Iroquois Theater in 1903, which killed over six hundred people, encouraged the City to consider exit routes and fire extinguishers. The Our Lady of Angels School Fire in 1958 prompted new code sections on noncombustible stairwells, alarm systems, and building addresses. And a deadly porch collapse in 2003 brought about stricter regulations for the construction of porches and exterior stair systems.

Today, Chicago's building code lists over 1,100 building code violations, ranging from nuisances and sanitation issues to structural defects.[37] This means that there are over a thousand ways a building could violate the building code. The most frequently recorded building code violations in Chicago are:

- insufficient heat (i.e., when a landlord does not heat buildings to 68 degrees over the winter)
- lack of smoke and carbon monoxide detectors (i.e., when a landlord does not provide these items for tenants)
- failure to maintain interior walls and ceilings free from cracks and holes (this, along with the following violations, can occur in rental and owner-occupied buildings)
- insects (this usually means cockroaches or bedbugs)
- issues with porch systems (this most often refers to wooden exterior stairs and decks)
- issues with windows (most commonly this refers to broken windows, but it could also refer to windows that do not open)
- issues with roofs, gutters, and downspouts (i.e., leaks, holes, and blockages)
- issues with exterior walls. (i.e., rotten wooden siding, crumbling masonry, or other structural damage).[38]

Chicago's building code demands discretionary decisions. Sixty percent of violations recorded by inspectors result from subjective decisions.[39] Some issues are more easily quantifiable, such as temperatures in rental units or whether a wall meets minimum size requirements, and can thus be evaluated with, say, a thermometer or a tape measure. More often, though, it is down to inspectors to decide what conditions amount to a violation. For example, should all holes in an interior wall count as a violation? What constitutes "a lack of sanitation"? How bad does a gutter blockage need to be before it becomes a violation? How many missing roof tiles does it take to count as a building violation? And when is peeling paint or overgrown weeds an issue? To be sure, as we will see, quantifiable issues also entail discretion.

The Buildings Department inspects all kinds of buildings, from historic single-family homes, slum-like rental buildings, and newly-constructed downtown high rises to churches, schools, and airports.[40] At the time of my research, the Buildings Department employed almost two hundred inspectors in twelve bureaus—ranging from elevator and refrigeration specialists to demolition and new construction permit inspectors.[41] This book focuses on inspections of existing residential buildings.[42] Although inspectors move between bureaus, there were between twenty-five and thirty residential inspectors at the time of my research.[43] Inspections of existing residential buildings are the most common type of inspections and have the broadest scope in terms of violations, ranging from peeling paint to caved-in roofs.[44] Residential inspections are follow-ups to service requests made through the 311 system.[45] In fact, residential inspections only occur in buildings listed in a 311 request; inspectors are not allowed to inspect properties without a service request as impetus.

311 Service Requests

Most large cities have city service request systems, commonly known as 311, where residents can file requests concerning issues such as potholes, streetlights, buildings, alleys, rodents, graffiti, abandoned vehicles, and trees that block sidewalks.[46] Chicago instituted its 311 system in 1999, to provide the public with a consolidated system to make requests for non-emergency city services.[47] Residents either call 311 or make a request online. Most buildings-related 311 requests come from tenants.[48] Others make them too, though, from neighbors and passersby to police officers, the fire department, family services, and local elected officials.

There is no formal system to prioritize or measure the seriousness of the hundreds of 311 requests that come in every week. Inspector John told me, for example, that

> [We] try and do the oldest one first. But it don't always happen that way because the one that says, you know, "falling down building" is put to the front [instead of] the one that says "my neighbor's fence is leaning on my porch." You know, it's not a perfect system, but it's the only system we have.

Supervisors direct inspectors to attend first to those requests where they deem that someone is in imminent danger. This can mean prioritizing newer requests over old, and is part of the reason the Buildings Department rarely meets the City's twenty-one-day deadline to respond to requests (and often has a backlog of over 5,000 overdue requests).[49]

Still, inspectors do eventually get around to following up requests.[50] From a call about bedbugs to calls about collapsed roofs, I observed inspectors doggedly investigating requests even when they suspected information was incorrect or out of date. "I can already tell you I'm going to find nothing," inspector Malcolm told me as he pulled up to a building in response to a call from a neighbor about a firepit. He begrudgingly got out of the car to take a look, though. Thus, inspectors do not exercise discretion in where they go and which buildings they inspect. Their discretion comes during and after inspections, as the following sections illuminate.

Heading Out on Inspections

"Ok, this is a judgment call," inspector Eddie said in a low voice, just out of earshot of the man whose condo he was inspecting. The condo unit was in a sorry state. Some rooms were scorched from flames from a fire a few days ago. The rooms that had escaped the fire were not much better. They were musty, and mold mushroomed across their walls and ceilings. The unit had no water and no electricity. "Basically, we're looking at there's really no water in the kitchen . . . and the floor is dangerous . . . the wiring is fucked from the fire and will need to be replaced. It sounds like [the owner] has a couple of options for temporary residency. . . ." Eddie trailed off as the condo owner approached, but I could see his mind working as he made notes on his clipboard. Eddie had to make the call as to whether this unit was fit for habitation, and whether the owner could stay in his home. This was Eddie's first inspection of the morning.

We had just come from the Buildings Department office—a lively place on a weekday morning. Inspectors, supervisors, and other department staff members occupy all four floors of a large red brick building on the Near West Side of Chicago. Most of the office space is taken up by cubicle desks, with supervisors' offices at the edges of these otherwise open floor plans. At the beginning of the day, the office is full of inspectors hunched over computers, collecting their 311 printouts for the day and entering inspection reports from the day before into the computer software. The office empties out around 9 a.m., as inspectors head to their cars to begin their day of inspections. This often takes longer than it should, as the office elevators are frequently broken and out of order—an irony not lost on inspectors, and a reality for which the department's elevator inspectors get a lot of good-humored hassle from their colleagues.

Inspectors begin a day of inspections with a pile of 311 printouts. The 311 system funnels requests that relate to buildings to Chicago's Buildings Department, where supervisors then allocate requests to inspectors, usually based on geographic proximity. Inspectors tend to work in either the north, south, or west of the city. Sometimes an inspector will "pick some 311s up" in a different area to help clear a backlog. In practice, however, the same inspectors usually do inspections in the same neighborhoods. This means inspectors get to know places, observe changes over time, and have a strong sense of what to expect in the neighborhoods they frequent. The areas—north, south, and west—are intentionally vast to help prevent familiarity between inspectors and the public that might give rise to bribery and corruption.

Corruption

There is good reason for the department to have policies pertaining to bribery and corruption; there is a history—both long and recent—of corruption in building departments in US cities. In Chicago, this history dates to at least the 1950s.[51] And in the most recent federal sweep—2008's Operation Crooked Code—the FBI indicted twenty-three Buildings Department employees for taking money in exchange for favorable inspections.[52] During my time at the Buildings Department, an inspector was arrested for taking a $300 bribe from a contractor. However, for numerous reasons, I do not believe this is the norm among the building inspectors I spent time with. First, inspectors expressed shock and fury when they heard of their colleague's arrest. "What an idiot," one inspector spat. "He will certainly go to prison, lose his job and pension . . . all for $300. He was fifty-five! Not worth it! Idiot, deserves jail and makes it harder for the rest of us." "Do you think most inspectors were surprised by this happening?" I asked another inspector the day the news broke. "Absolutely," he responded without missing a beat. "Shocked would be more appropriate. Disappointment followed by dread. Dealing with the public is hard enough. When they think you're corrupt, it's even more difficult." Inspectors believed that their bribe-taking colleague was foolish for risking his job and the pension he was so close to receiving, and they were angry that they would suffer the reputational consequences in a profession already stigmatized for corruption.

Second, most contemporary corruption cases involve inspectors who sign off on new construction and permits,[53] not the inspectors who follow up on 311 requests (who feature in this book). The latter often do not meet people (and therefore lack the opportunity for bribery) and rarely have power to either thwart or grease the wheels for a lucrative new development. Third, as we will see, cutting corners and unfair profit are traits that inspectors disdain. Inspectors also regularly bemoaned corruption in other contexts: from city officials in Flint, Michigan, and Chicago's political elite, to the banking and mortgage industry, and condo boards.

From Request to Inspection

While corruption occurs, inspectors have other things on their minds: trying to clear the perpetual backlog of 311 service requests. Each 311 printout lists information provided by the requester, either taken down by a 311 operator during a phone call or submitted on an online form: whether the issue is life-threatening, whether the building owner knows about the issue, the precise location of the issue, whether the building is owned by the Chicago Housing Authority, whether the building is owner-occupied, and contact information for the requester. Much of this information is usually missing in most reports. One inspector estimated that 311 requests provide no more than a description of the issue as much as 70 percent of the time. This means that inspectors typically work from 311 printouts that only state an address and something like "hole in roof" or "bedbugs." People seldom leave contact information.

Although inspectors can access an online system to look up the inspection history of buildings before they go out into the field, this rarely occurs. Inspectors do not have time to do this in addition to going on inspections and filing reports. So, while they may have expectations based on what they know about the surrounding neighborhood, inspectors usually turn up at an address with little to no knowledge of the building in question. Inspectors also do not schedule inspections ahead of time. They knock on doors and ring doorbells, but, because regular inspections occur on weekdays between 9 a.m. and 5 p.m., there is often no one home to let inspectors into buildings. This means that inspections typically entail the inspector walking around the exterior of a building, looking at porch systems, windows, brickwork, roofs, and yards by themselves.

Inspections can last ten minutes or take up most of a work day, depending on the size of a building and whether it is possible to access the interior or only to view the building from the street. For these same reasons, an inspector can complete only one or as many as ten inspections in a day. Whether inside or outside, an inspector's job is to assess the building for code violations. Inspectors usually record multiple violations per building per inspection. Although the mean number of violations per inspection in 2015 was eight, some inspections result in over thirty recorded violations.[54] Building violations are listed online, in the City's Data Portal. Chicago's Data Portal originated in 2012. Currently, the system holds over 13,000 data sets, ranging from locations of stolen bikes and business licenses to building permits and building violations. As with most other large US cities, records of building violations are thus visible online through the Data Portal.[55] Property owners, tenants, developers, buyers, sellers, city officials, and the media can easily see the violation history of a building, and whether violations have been marked as resolved.

Making Decisions

It's late afternoon, and laughter punctuates concentration and near silence as someone cracks a joke, shouts about the game last night, or swears loudly at the computer after entering something incorrectly into the department software. The computer system is the nemesis of most inspectors, and some reminisce about when there were staff people in the Buildings Department tasked with entering reports (or, in inspectors' words, when there were "girls to do the paperwork"). By 3 p.m., many inspectors have finished up their inspections for the day and have returned to the office to file their reports.

Inspectors have decisions to make. They must enter one of four options into their computer software: no action, a notice, an administrative hearing, or a building court case. These are shown in figure 6. While there are no city-wide data on the frequency of the various outcomes of inspections, my methodological appendix provides this information for one neighborhood in Chicago. The following sections offer an overview of these four courses of action and how each affords inspectors the discretion to exercise leniency or punishment at various moments (see table 1).

No Action and Notices

Inspectors can select "no action" if they do not find code violations. This rarely happens, however, because inspectors can always find issues with buildings, and little separates no action from a notice. A notice is the modal course of action when violations are few and relatively minor, such as peeling paint, a cracked window, or overgrown weeds. A notice prompts

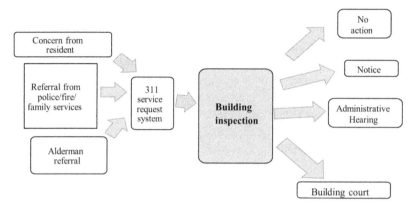

FIGURE 6. The origins and outcomes of residential building inspections.

TABLE 1 **Inspectors' Points of Discretion**

	Leniency	Punishment
During inspections	Only writing up severe violations	Writing up every violation
After inspections	Notifying owner, but no other action; recommending building court (knowing this provides opportunities for leniency)	Sending building to administrative hearing
In building court	Giving ample time to comply; dismissing case early; not adding additional violations; not exacting fines or fees; not insisting on retroactive permits	Giving little time to comply; adding additional violations; levying fines and fees; insisting on permits

a letter to the owner of the building in question, notifying the property owner that they need to make repairs and arrange a follow-up inspection. But owners are not given a deadline to resolve violations, and the City does not follow up on these notices. The onus is on the building owner, and the City has arguably fulfilled its duty by notifying the owners of issues. Notices are thus a way for the City to avoid culpability in the event of an incident. In addition to this broader political context, a notice is one way inspectors show leniency to property owners, because this action does not force owners to make repairs, which can be costly, and does not ensnare them in legal action.

Administrative Hearings

If a property has numerous but minor violations—ranging from issues such as broken windows to a chimney that needs tuck-pointing—an inspector may opt to send a property owner to an administrative hearing. Administrative hearings—which also cover traffic violations and police tickets for misdemeanors—are quasi-judicial proceedings, involving an administrative law judge, defendants, and their defense attorneys. Administrative law judges have no legal power to order a property owner to resolve issues. Instead, property owners are ordered to pay fines for violations, which can range from $200 to $500 per violation, per day, plus $75 in court costs. On top of the bureaucratic red tape and time spent at the hearing, this can be a costly outcome for a property owner. Between 2006 and 2015, there were 73,843 building violation administrative hearing cases in Chicago. The average fine was $1,575.[56] The City collects these fines as revenue.

Although the risk of further fines may encourage owners to resolve violations, hearings officers seem to care little about fixing issues, demanding fines irrespective of any work a property owner has done between the inspection and the date of the hearing. Property owners and their attorneys rarely successfully contest an inspector's report or avoid fines. Inspector Micky told me that the "problem is that fines do not equal compliance. We used to call it 'death by a thousand cuts,' but there are just not enough building inspectors anymore." He meant that the system used to be more effective because there were more inspectors to record violations and send buildings to a hearing. Inspectors view administrative hearings as a money-making "fine machine" for the city, a confusing bureaucratic process, and an inefficient way of improving substandard housing conditions. While property owners receive fines when they are sent to administrative hearings, this does nothing to fix building violations. Yet, administrative hearings remain a course of action inspectors commonly choose when they want to punish property owners or when issues are not serious enough for building court.[57]

Building Court

In contrast to administrative hearings, the focus in building court is on resolution, and judges have power to order code enforcement. Cases in building court last for months and sometimes years, as property owners gradually make repairs (or fail to). Judges also regularly do not fine property owners if they are making progress toward compliance. Taking up the eleventh floor of Chicago's downtown Circuit Courthouse, building court groups building violation cases into eleven court calls, ranging from specific calls focused only on heat requests or exterior walls, to more general calls for "occupied buildings with general code violations," which is divided into three court rooms by geographic location (north, west, and south). Each court call has a corresponding judge, city attorneys, and a building inspector. Court inspectors track the progress of repair work on buildings by scheduling an inspection a few days before the property owner is next due in court.[58]

In court, inspectors give testimony as expert witnesses, testify about the condition of buildings, and provide progress reports. Judges rarely question their expertise or testimony. They have inspected the building, and they have the technical expertise and photographs on hand to prove any point they are making about the condition of a building. Although attorneys and defendants get to make their case in building court (unlike in an administrative hearing), inspectors' accounts are considered the gold standard. Property owners' attempts to dispute inspectors' reports were never (that I witnessed) successful. Judges always sided with inspectors. Inspectors are supposed to refer owners to building court if they deem that a property has serious safety issues (building owners that have multiple administrative hearings cases within one year are also sent to building court). Supervisors teach inspectors that some issues—porches, illegal conversions, a lack of smoke and carbon monoxide detectors, and any other "dangerous or hazardous condition"—must go to building court. Inspector Danny told me about an inspection he had done the week before, following up on a request about an unstable back porch.

> It was fucking raining out and I was hoping against hope I was going to find nothing. But I come around the back and there's a shitty porch. And the woman who complained on the second floor wasn't home. So the owner comes out: "what do you want?" I show him the complaint[59] and he said, "I already fixed it." And I say, "you're not quite there and I can't just walk away." "So you're going to fuck me?" those were his words. "No no, I'm not going to. . . ."

Danny seemed upset as he recounted this story. The owner did not grasp the professional obligation Danny was under to send the building to court. Danny was following the rules. He also did not want to be held responsible if anything happened at the property after he had inspected it.

"I can't not write a building once I've seen something dangerous," he said, "I'd be the last guy there." While the definition of dangerous is subjective, an inspector is more likely to err on the side of caution when it comes to porches because of high-profile fatal collapses. Inspector Nick, for example, told me that a porch "doesn't have to be dangerous, just not up to code and you'll be in court." I frequently heard discussions in the court room where the definition of dangerous was up for debate.

While certain issues may land a property owner in court, inspectors can still exercise discretion during the process. Inspectors make recommendations to the judge based on progress inspections before each court call. So, while a dangerous porch leads to building court, the court inspector assigned to the case has a lot of discretion to decide—most often without any pushback from judges or city attorneys—how long an owner has to make repairs. An inspector can show benevolence by recommending an ample period of time in which building owners must fix violations. An extra three to six months can make a significant difference to someone who is trying to find the tens of thousands of dollars necessary to fix a porch. On the other hand, the inspector can suggest that an owner be given little time to fix issues, or can add violations from follow-up inspections to an ongoing court case. This can add to the property owner's headaches, piling up more court dates and more paperwork. Inspectors also decide when a building owner has made sufficient progress for the court case to be dismissed. Oftentimes building owners do not remedy *every* violation before cases are dismissed, again affording inspectors discretion. While building court is characterized by leniency toward defendants (at least in contrast to eviction courts, for example), inspectors still make decisions at various points in building court than can amount to leniency or stringency.

All frontline work is characterized by discretion. Laws, policies, and regulations have subjectivity written in and are rarely interpreted or enforced uniformly. In most frontline settings—from the police to social work—workers enact discretion by deciding who deserves leniency or punishment. This chapter has detailed the particular moments, locations, and opportunities in which building inspectors enact their discretion. Leniency during inspections can mean overlooking minor violations— broken windows and peeling paint, for example—and not recording them. Punishment means nitpicking: taking a fine-tooth comb approach and recording every possible violation, including such potentially trivial issues as overgrown weeds. If an inspector does record a violation, however, there are still opportunities for compassion further down the line. An inspector can record violations but not send the building to court. In such a case, the inspector issues a notice to the property owner, but takes no additional action. This means that the building owner is not mandated to resolve issues. Conversely, the inspector could opt to send the building owner to an administrative hearing or building court. Building court inspectors can be lenient by allowing ample time to fix violations and not recommending fines or fees when the case is dismissed. Inspectors can also recommend that judges dismiss property owners from building court before all issues are resolved and without property owners obtaining permits for repair work. In contrast, punishment entails recommending little time to make repairs, adding more violations, levying fines and fees, and insisting that owners pay and wait for permits for repair work. In short, code enforcement is rife with opportunities for leniency or punishment.

How Stacked Decks Shape Frontline Work

Building code enforcement in Chicago is a particular case of frontline work, but we can use it to sketch a framework for how stacked decks might shape frontline work in other settings. To do so, it is useful to ask what is common or unique about Chicago, code enforcement, and building inspectors. Some aspects of Chicago's stacked deck pervade other US cities: racial and economic inequality coupled with the lack of affordable housing, the financialization of rental housing, aging housing stock, and rising costs of construction, for example. But racial segregation is starker in Chicago than in many other cities. And because it is an old city, broad national trends of disparities are magnified and "poverty and wealth confront one another in the next block or across the street."[60] As such, Chicago and other segregated, old cities may offer their frontline workers the most explicit manifestation of the stacked deck. Thus, inspectors in Chicago may make sense of the stacked deck in much the same way as inspectors in Philadelphia, for example, unlike inspectors in Miami. There are also likely to be city-specific characteristics that shape inspectors' assessment of the stacked deck. The context of post-Katrina New Orleans amid widespread building damage, for example, may have expanded inspectors' categorizations of precarity and prompted them to show leniency to a broader group of residents than before the storm.

If buildings are especially expressive of disparity, inspectors may share stances with other professionals who deal with housing, such as realtors, appraisers, and assessors. While not frontline workers, these people still make decisions with great significance for urban populations and landscapes. Yet, a recent investigation and action against the Cook County Tax Assessor's Office tells a different story. The investigation illuminates that, like code enforcement, property assessments are subjective and discretionary. Unlike their counterparts in the Buildings Department, however, property tax assessors allegedly consistently go easy on wealthy homeowners and excessively tax homes in minority communities—"shafting the little guy" through regressive valuations.[61] Their lack of face-to-face interaction with the buildings they assess may help to explain these attitudes and actions that seem so divergent from those of inspectors. Property tax assessors' evaluations might change—and come to resemble more closely those of building inspectors—if they were to come out from behind their desks and visit the properties they assessed.

There are a multitude of other occupational characteristics that shape a person's recourses and capacities to act. Jobs with more or less discretion, and more or less oversight, might prompt people to see and categorize stacked decks differently. Recent investigations of traffic tickets in Chicago and fines and forfeitures in California, for example, reveal the uneven racial geography of ticket debt and revenue extraction in communities of color.[62] Yet, in the absence of empirical attention, we do not know how much discretion ticketing entails, or what discretionary decisions look like. There may be differences in tickets for parking violations versus tickets for outdated city stickers. If discretion permitted, perhaps police officers would cite a flashy car with outdated stickers more readily than a modest car, following the logic of building inspectors' decisions. Or perhaps they do not because police officers have less autonomy than inspectors, or their mobile technology tracks and regulates decisions in a way that automates discretion and thereby reproduces inequality.[63] What becomes clear is that there is a great deal more we could know about how social locations, recourses and clout, and experiences with stacked decks shape the institutionalized practices of other frontline workers.

Conclusion

Look around you. Does the building you are in, where you work or where you live, have more than one external door? Does it have a fire escape? Are the windows big enough to let in a good amount of light and air? Though we might not realize it, chances are that we have all benefited from building code regulations. Most people live their lives inside buildings, yet we rarely think about the construction materials, egress options, or stability of the walls around us. This book highlights the importance of mundane aspects of city life. Fire codes, nuisance ordinances, zoning, parking restrictions, building standards: the regulations and laws that govern people, places, and property in cities have always been selectively enforced.[64] We are familiar now with how government regulations have reinforced inequality in our cities, from zoning decisions to redlining;[65] yet the enforcement of building code regulations has received less attention.[66] This chapter has explained what motivates inspectors to selectively enforce the building code, and how the framework of code enforcement affords moments of discretion that enable them to do so. In subsequent chapters we will see how these moments of discretion unfold across the city—from new luxury apartments to vacant and foreclosed properties, and everything in between. We see inspectors using the disparities they see every day to inform their decisions about leniency and punishment—or, as I call them, their stabs at justice.

CHAPTER THREE

Rentals and Relative Assessments

The noise of the buzzer set dogs barking somewhere in the building, but no one answered the door. He only gave it a few seconds before he pressed another. There were twelve in total, and building inspector Malcolm must have pressed them all within thirty impatient seconds. I was standing next to Malcolm on the porch of a twelve-unit apartment building he was trying to get inside to inspect. Although Chicago's North Side is majority White, this area of the far North Side is one of the city's most ethnically and economically diverse neighborhoods, home to large communities of Hasidic Jews, Indians, Pakistanis, and Koreans.[1] Low-rise apartment buildings and large single-family homes mingle on neighborhood streets that are peppered with restaurants and small food markets that reflect the area's diversity.

Malcolm, who wore his blue Buildings Department polo tucked into his khakis, became an inspector almost fifteen years ago, after his family's contracting business went bankrupt. He is White and working-class, and he takes pride in his Polish ancestry. We were following up on a request alleging that the landlord had padlocked the fire escape and had done nothing to abate a months-long bedbug infestation. But because no one was answering the door, we could not get access to see if the fire escape was padlocked, and we could not be sure there were bedbugs. Malcolm had given up—no one was home. Just as we turned around to leave, however, a young woman approached the doorway, taking keys out of her bag. The woman, who was Black and dressed in faded jeans, seemed hesitant to talk to Malcolm at first, but nodded when he asked about the bedbugs and padlocked back gate.[2]

Her confirmation appeared to spur Malcolm on. He took a few steps backward and craned his neck to scour the area he could see from the front of the building. He had found something to write up. It only took a few seconds for Malcolm to point to a mass of loose wires hanging at shoulder height from the exterior wall just around the side of the building. "I wouldn't normally write this up," he told me as he scribbled. Yet he did. He would enter the violation for dangerous wiring into the computer system when he got back to the office. I couldn't help but notice the determined look on his face. It seemed, I thought, like he had been looking for something to use to punish the landlord of this rental building.

Do other building inspectors act in the same way? Do they all punish landlords? To get to the bottom of this, I did some digging and pieced together building violation records, 311 requests, and information about building characteristics across the city. Sure enough, areas with more rental buildings have consistently received more building code violations per 311 request than areas with fewer rentals and more owner-occupied housing. This means that, when they follow up on a 311 request, inspectors like Malcolm are more likely to record violations in areas with a lot of rental buildings. And this is not because rental buildings tend to be bigger or older.[3] It is because inspectors want to hold rental properties accountable, so they make punitive stabs at justice — small-scale attempts to ameliorate or make up for disparities between profit and precarity. When they inspect rental properties, they often issue building violation citations about small things, and in large quantities. But not all rental buildings get the same treatment. Sometimes inspectors' stabs at justice entail letting rental properties off the hook for code violations.

"You know, this really isn't in such bad shape," inspector Bill said, kicking broken glass and peeled paint chips across the water-damaged carpet that was once pink. I wondered who he was trying to convince. Bill is one of Malcolm's colleagues. He is White, and nearing retirement. Like Malcolm, inspector Bill began his career in construction. He still wears his carpenters' union jacket over his Buildings Department T-shirt. On this afternoon, we were responding to a request about the condition of a three-flat apartment building that had been vacant for some time. The vacancy is not too surprising considering the surrounding neighborhood, which has undergone significant depopulation since the 1960s.[4] But Bill did not need any tenants to tell him about the condition of the building, which had more broken than intact windows. The front door was permanently open, missing a lock and handle, and water had damaged portions of the floors. Bill warned me to watch my step, fearing that I would step on the broken glass that littered the floors beneath the windows. He could have cited this property owner for each of these issues—they are all violations of Chicago's building code, meaning potential fines or court cases for the property owner. But Bill decided only to give the owner of this rental property a notice, which amounts to mailing a notification letter to the owner. Notices do not mandate repairs or even a follow-up inspection.

Bill's decision flies in the face of a lot of existing academic research on urban conditions and inequality. Studies demonstrate that we are all more likely to perceive dilapidation and disrepair in low-income and minority communities.[5] While Bill's decision was not punitive, other studies suggest that frontline workers like Bill—who, like most other inspectors, is White and working-class—would heavily penalize the owner of this building, located in an African American neighborhood with a high poverty rate.[6] This is, after all, the conclusion of many studies of broken-windows–style policing. To be sure, as we drove to the inspection, Bill told me that the neighborhood was in "bad shape," and pointed to apartment buildings on neighboring streets that were the subject of current court battles and were facing heavy fines for building violations. Yet Bill seemed to be doing the opposite: *not* policing these broken windows.

Why is it that Malcolm decided to punish a rental property, but Bill did not? It is not because the first property was in worse condition than the second. And it is not because each inspector makes their own, often dissimilar discretionary decisions. As this chapter demonstrates, inspectors' decisions are not motivated solely by their assessment of the physical condition of buildings, and there is a great deal of consensus (due to the shared workplace culture, informal training, and working-class backgrounds we read about in chapter 2) among Malcolm, Bill, and their colleagues when they make decisions. Instead, the divergence between Malcolm's and Bill's decisions is because the buildings they inspected occupy different categories in inspectors' minds. Malcolm came down hard on a landlord of a relatively large building, who was making rent while neglecting to fix issues that were endangering his tenants. Bill went easy on a landlord of a small building in a disinvested neighborhood, who was not collecting rent.

As this chapter reveals, inspectors' perception of the stacked deck provides them with clear cues about how to categorize buildings and how they should direct their stabs at justice. We see inspectors making distinctions about negligence and scales of profit, as well as the size of buildings. We also learn how inspectors evaluate effort and struggle. These are all *relative assessments* that prompt inspectors to penalize big, professional companies, slumlords, and corner-cutters, and to try to protect small-time landlords and family buildings.[7] We also see inspectors making distinctions between deserving and undeserving tenants. Inspectors' categorizations of people and places may not always be accurate, we find out, but if we pay attention to the logic that underlies them, we can uncover just how essential relative assessments are for frontline workers as they do their jobs and try to shape the city.

Negligence and Disproportionate Profit

Negligence is inspectors' primary concern when it comes to rental buildings. But how do they decide what constitutes negligence? This section demonstrates who counts as a bad landlord and the role of disproportionate profit and corner-cutting in inspectors' decision-making. Inspectors frequently assume that violations in some rental buildings are due to willful negligence—or malign neglect, as I call it—and try to penalize landlords accordingly.

Bad Landlords

"We're only going after the people that are bad landlords," inspector Dave announced with what sounded like pride. We were sitting across a desk from each other in the Buildings Department, and I had just asked him to explain his job. Dave's comment about bad landlords echoed what other inspectors had told me, and with similar zest. But what counts as a bad landlord? What makes a landlord bad or not bad, in the eyes of these frontline workers who see all kinds of buildings and all kinds of landlords? For inspectors, a lot comes down to profit and proportion. Inspectors scorn landlords who do not maintain their buildings while continuing to collect rents. The worse the condition, or the higher the rent, the greater their scorn. "We should jail slumlords!" one inspector told me. "And I mean prison." While the image of a bad landlord conjures up crumbling buildings with broken windows and rats, inspectors do not come down hard only on owners of these types of buildings. They willingly penalize landlords of all kinds of buildings who violate the code, spanning the spectrum of downtown high rises and new luxury rentals in gentrifying areas to dilapidated buildings in high-poverty neighborhoods.

On an inspection of a six-unit apartment building in a North Side neighborhood, Nick—a White inspector with thinning hair and a greying moustache—pointed to the accumulation of trash and dried mud lining the carpet in the stairwell. "This is a bad building owner," inspector Nick said, because "there is no maintenance guy coming by to clean every few weeks." The son of Italian immigrants, Nick grew up in a White, middle-income Catholic parish on the far South Side of the city that is now majority Black. Like many inspectors, he used to work construction and has worked at the Buildings Department for well over twenty years. He lives in a small bungalow that he shares with a cat he found four years ago in a vacant building.

Inspector Nick probably felt at home in the neighborhood we were in. The area is majority White (with a small Latinx population) and home to streets and streets of bungalows interspersed with small apartment buildings. It was one of only a handful of Chicago neighborhoods to gain in population between 1980 and 1990, and it has a reputation as a stable community of homeowners.[8] I saw Nick grinning sheepishly as we looked at the porch around the back of the building. "It doesn't look good," he said, maintaining his smile, "and the owner won't be able to upgrade it. It will need to be replaced. And it's not going to be cheap." Nick seemed to relish his power to use building violations to incur costs for the property owner because they were not keeping the building clean. This was a stab at justice, an attempt to take this "bad building owner" down a peg or two. Nick was responding to a relatively small sign of negligence: a landlord not keeping their building clean. But inspectors come down especially hard when they believe landlords and property owners profit from extreme disrepair.

Praying for Landlord Hell

Inspector Eddie wiped his forehead with the back of his hand. It was a hot and humid afternoon and there was no shade. That there were no big old trees to offer shade in this neighborhood is not incidental. Neighborhoods—like this part of Chicago's Southwest Side—that have been the site of disinvestment followed by predatory lending and foreclosure have fewer trees and are hotter, on average, than wealthy neighborhoods.[9] Today, the neighborhood is notorious among inspectors and the public for its empty buildings, violent crime, poverty, and vacant lots. Due to the legacy of segregation, the neighborhood is over 99 percent African American. As we walked up to the six-flat, a man approached us to tell us he was the one that made the initial request about a lack of electricity and cockroaches, and he led us inside his apartment, where he lived with his wife. The couple were both Black and seemingly in their twenties, and both wore jeans and T-shirts. They occasionally talked over each other as they went from room to room pointing out numerous violations throughout their apartment that the landlord had refused to fix, ranging from holes in the walls and clogged drains and sparking outlets, to kitchen and bathroom ceilings that were thick with orange mold, and their neighbor's trash and mop bucket tipped across their back doorway. The back porch was unstable, the back gate was padlocked, and various insects clung boldly in plain sight on the kitchen walls and floors. The landlord had also failed to provide heat last winter, or electricity for the past three months—this was the height of the summer, but the use of air conditioning units, or even fans, was impossible. The absence of electricity had also meant the tenant was unable to use some medical machines that he needed, giving him little choice but to readmit himself to the hospital. His insurance company took back the machines in response.

The couple were disgusted with the condition of their apartment and their landlord's negligence. "See that?" the woman asked, barely able to keep herself from yelling while pointing to the bathroom ceiling. "She just don't care. At all." Inspector Eddie seemed disgusted too; he had been furiously scribbling notes the whole time, telling them he was sending the landlord to court and he wanted to get every detail down. As we were leaving, Eddie asked them what they paid for rent. "I ask," he explained to them, "because when I pray for landlords to go to landlord hell, it's based on how much rent they charge for a shithole." "Yeah, she's on her way to landlord hell," Eddie said after the couple told him their rent was $750. Eddie's disdain was rooted in the negligence of this landlord coupled with the rent she was charging. To Eddie the rent equaled unfair profit.

The write-up for this building listed numerous serious violations, and Eddie sent the owner to building court. At the minimum, the owner would be forced to fix the porches, electricity, and ceilings. After the inspection, I wondered what motivated Eddie's reaction. I knew a little about his working-class background, and I thought that had something to do with it. Eddie is White and grew up in a South Side Irish Catholic neighborhood as many Whites were leaving for the suburbs. He completed a year of college but did not get his degree before becoming a construction foreman. Like his colleagues, Eddie disdains exploitation. To be sure, the landlord's profit margins may not have been as wide as Eddie assumed. For all he knew the owner of this property was financially incapable of making repairs. Yet, he had classified this as a case of a bad landlord and made a stab at justice accordingly.

You might be asking yourself why anyone would rent a unit in this building, considering the conditions. The answer is that often low-income tenants have little to no choice. The prevalence of slumlordism, coupled with limited government regulation, means that renters at the bottom end of the housing market have few options. Units are poor quality, repairs go unaddressed, and landlords get away with charging rents that are not that different from those in well-maintained buildings.[10] There are even fewer options for renters of color due to racial discrimination, for renters who have experienced evictions, and for families with children.[11] And landlords can exploit the fact that these populations have so few options by charging high rents for slum-like conditions.

I was interested in how much inspectors knew about the landlords they penalize. In a majority African American neighborhood one day, I asked inspector Nick if, based on the location, he could guess the race of the landlord whose building he had just written up for deplorable living conditions. He paused to think before he answered: "you know, back in the day the quintessential slumlord was a White guy with Black residents, and that's not so much the case anymore. . . . And, you know," he continued, "I was initially surprised when I first started [on the job] by things like Hispanic slumlords of Hispanic tenants. [But] it's not really about Black or White landlords, it's about land-*lord*." Nick's resentment toward slumlords was not rooted in the marginalization of tenants, or a concern for racial discrimination, but rather in the power of the landlord and the class of rentiers (people who profit from an asset such as housing) who make money without earning it.

Inspectors' assessments about disproportionate profit—from Eddie's ideas about who warrants a place in landlord hell to Nick and Dave's ire at unmaintained rental properties and their owners—may not always be accurate. But they tell us that calculations about profit and proportion are tools inspectors use to categorize the city and to take action. These kinds of categorizations are not peculiar to high-poverty neighborhoods and squalid rentals. As the next section shows, inspectors find bad landlords all over the city, renting out all kinds of apartments.

Corner-Cutting

As Chicago neighborhoods go, this one—a Northwest Side neighborhood known for its historic bungalows—is diverse. While Whites constitute the majority of the population, almost half of the population is foreign-born, and there are sizable numbers of Black, Latinx, and Asian residents. Inspector

Nick and I had just managed to get through the door of the two-story, vacant brick building before it started to rain. The building was in court and Nick needed to check whether the owner had cleaned up the mold in the basement or worked on the rickety porch. The basement had been gutted and refinished. But that was not all. The building also had new electrics, a new furnace, and new stairs. In Nick's eyes, this was great; rehabs nearly always take care of existing violations in a building. His mood changed, however, when he checked the permit that was displayed—as a legal requirement—in the window at the front of the building. The permit covered 1,100 square feet of drywall, a new bathroom, and other minimal repairs. The permit did not cover the gut rehab that was underway.

"You might want to stop working at this point," Nick called over to the workmen, "[the owner] is going to have to get a new permit." He later explained that he had not told them they must stop work, nor put an official stop-work order on the building, because he wanted to make the owner "learn the hard way." Not only would the owner now have to remove all the drywall in the building just to have an inspector check the electric wiring behind it, and then replace it, he would also have to continue to pay the workmen if they kept working. Nick reasoned that "they probably have kids to feed and this way the owner will still have to pay them." He seemed to enjoy the power he had to penalize the owner, who had tried to pull the wool over Nick's eyes by displaying an inadequate permit.

I was still taking notes in the car after the inspection, when he eagerly told me that something else had helped him make his decision: the "finish" of the kitchen. The granite countertops and the expensive chrome appliances, he told me, signaled to him that the owner was planning on flipping the property and would thus be passing the insufficient permit issues on to the next owner. Maybe Nick also knew that, over the past few decades, neighborhood officials had been encouraging development and upscaling[12]—precisely along the lines of this building renovation. Nick opted for a heavy penalty because he thought the landlord was trying to profit by taking shortcuts. But Nick also hoped to protect the workers and the future owner of the building. In this context, negligence is rooted in corner-cutting rather than dilapidation, and stabs at justice are especially retributive.

Malign Neglect

To Nick and his colleagues, corner-cutting in buildings owned by people with resources signifies an intentional disregard for repairs or maintenance. This classification derives from the disproportionate relationship inspectors infer between the owner's profit margins and the condition of buildings. As such, a building that conveys luxury and contains code violations—ranging from an odd broken window or shoddy workmanship to complete disrepair—signals intentional negligence, or as I call it, malign neglect.[13]

Inspector Bill made just such an assessment on a sunny but cold afternoon on an inspection of a rehabbed four story, forty-plus-unit apartment building on the main commercial strip of a neighborhood on the city's North Side. Though historically ethnically diverse, this part of the city is changing rapidly and becoming increasingly White. Other telltale signs of neighborhood change abound nearby. The building sits opposite a CrossFit studio and a restaurant selling small plates of Mexican "street food" for over $15, and there is a new Whole Foods less than a mile away. The building—advertised as luxury-style apartments—exemplified other rehabs in the area and boasted a shared roof deck with a view of Chicago's skyline, a boutique hotel–like lobby, a high-tech video surveillance and entry system, a dog run, covered parking, sliding barn-style doors, spa-style baths, and high-end finishes.

The luxury look of these apartments is not the whole story. To maximize the number of units in the building, developers had squeezed bedrooms into the corners of floorplans. As a result, the bedrooms had no external windows—the only source of light and ventilation were cutouts in the bedrooms' partition walls. These cutouts did not meet code—some were covered by blinds and others were too small. Bill expressed his annoyance as he kept finding the same issue as he careened irritably from one apartment to another throughout the large building, with me trying to keep up. Shaking his head, he told me this kind of building was all the rage with developers on the North Side, who were converting old buildings into micro-apartments but still charging rents comparable to those for larger units.[14] "300 square feet for $1,000 a month?" Bill said incredulously. "That's what shouldn't be allowed in the code!" Other inspectors share Bill's annoyance with the assumed profit margins of developments like this. Bill was annoyed by the corner-cutting, but something else had also prompted his ire: the size of the building.

Sizing Up Rentals

In this section, we see that inspectors use size as another relative evaluative tool to assess buildings. Big multi-unit buildings convey to inspectors that landlords are professionals and should be held accountable for violations. By contrast, inspectors see small rental buildings as a precarious category of housing and seek to protect them. Their sympathies run out, however, when small-time landlords negatively affect their neighbors.

Big Buildings

Rental buildings vary in size. In many of Chicago's neighborhoods, two- and three-flat rental buildings predominate. In others, large buildings house ten, twenty, or fifty units, sometimes more. Even single-family homes can be rental properties.[15] Each of these building sizes is meaningful to inspectors. So meaningful, in fact, that city-wide trends in building violations fall along these lines. I went back to the data set I had compiled (of building violation data, 311 requests, and information about building characteristics). Analysis of the data set shows that there are more building violations in areas with more rental buildings with over six units than in areas with small rental buildings.[16] This pattern holds even if we account for the number of 311 requests in the area. It seems inspectors give out more violations to owners of larger rental buildings than to those who own small buildings. But this trend is not because larger buildings have more issues by virtue of their size. Inspectors only cite a property once for an issue even if it occurs throughout the building. For example, even if smoke alarms were missing in multiple places, the property would only receive one violation for missing smoke alarms. Instead, the pattern can be explained by inspectors' assumptions about proportion and profit.

Revenue from a rental building signals to inspectors that the owner has funds to maintain the property. To inspectors, big buildings signal professional landlords—someone who is in the business of renting property solely for profit. Paul—a White inspector with almost thirty years experience—concurred. He laughed as he told me that "if there's a corporation that owns a whole bunch of buildings . . . and they have issues, then yeah, I'm going to hold them to the very small violations." Inspectors use the terms "big buildings" and "professional" landlords interchangeably. To them, professional landlords who manage large properties and multiple buildings deserve their scorn.

Inspector Frank agreed. "If it's a multi-unit [building] and if there's twenty-seven violations, I won't give them any breaks." "Hey, I'm not doing it because I'm trying to fund my pension," he stressed cautiously as he looked around the downtown coffee shop where I interviewed him, seemingly concerned about who might be within earshot. "I'm doing it because I know they can afford to pay. And they *should* pay!" Inspector Nick stressed the relative nature of his logic, explaining that he thought it was important to "look at an apartment building that generates revenue" in a different light than he would single-family homes, which he described as "a sole investment that doesn't generate revenue." Nick continued:

> There's a replenishment fund that is or is not being used to maintain the property. So, I think when you hit the landlord of a six-unit, who can afford to fix it, or if he needs to go the bank, the bank's more inclined to give him a loan because they have an income-generating property. To me it's about equity.

Nick failed to mention that single-family homes can also generate revenue when their owners rent them out.[17] Similarly, big buildings do not always generate profit. Inspectors may not always be accurate in their distinctions between tenure types and thus in their categorizations of profit. Yet size is a tool of evaluation for inspectors when they have little other information about the properties they inspect. Building size signals to inspectors whether owners have funds to maintain their property.

Inspectors see big or professional landlords as particularly problematic in neighborhoods they assume are lucrative because of prior disinvestment. The radio was turned down as we sat in inspector Eddie's car one afternoon, surrounded by boarded-up houses. We were between inspections in a South Side neighborhood that had been hit hard by foreclosures. The building on the corner stood out, however, because it was mid-renovation. Scaffolding stretched across one corner and the windows were so obviously new that they still bore brand-name stickers. "Nobody's in this neighborhood magnanimously," Eddie told me. "It's for profit and to generate interest in the development community." "The federal money that goes to it, it's for profit too," he added, before launching into an imagined conversation between HUD and a real estate investor. "Hey, will you take over half a million apartment buildings in Chicago that used to be HUD controlled?" "Sure, what's in it for me?" asks the make-believe investor. "Huge sums of money." Eddie continued:

> One guy, he owns 100,000 units in Chicago, maybe more. You see it a lot. Investors come in because you have a 36- or 48-unit building. You're getting federal money for construction loans, low-interest construction loans, if not outright grants, to take a property you're buying for next to nothing . . . now you can afford to put that million bucks into that 48-unit building. A million dollars is

a lot of money and they're beautiful. And because they know their rental base is Section 8, [they're] guaranteed [that] income, and way above market rate. I don't know where they come up with these fucking calculations. Market rate here for a shitbanger two-bedroom is probably $800-1,000. A renovated two-bedroom unit is $1,400, a three-bedroom, $1,600-1,800. You can do the math.

Several things about this situation enraged Eddie: that investors amass huge dilapidated "shitbanger" rental portfolios, that they are savvy enough to utilize government incentives for profit, and that they rent units out to housing choice voucher recipients—who have fewer options on the rental market—to get extra money. He used what he knew about the financialized low-income rental housing market to deduce that a profiteer was stacking the deck.

Small-Time Vulnerability

Inspectors do not see all landlords as profiteers, however. In fact, they believe that some landlords make little—or even no—profit from their property because the deck is stacked against them. This is particularly problematic to inspectors because they see small-time landlords working hard against the odds. To inspectors, this is most often the case for two- and three-flats owned by small-time landlords (sometimes called mom and pop buildings), and "family buildings"—a term inspectors use to mean a two- or three-flat owned by someone who rents the other units to family members. I asked inspector Malcolm about small rental property owners, and he said, "if I had to generalize, I'd say they were the naïve type ... who think they can make money from rents but don't have the overhead [to cover maintenance and issues that arise]." Inspectors share the idea that small-time landlords make very little in the way of profit, and that any profit they do make is virtuous relative to the unfair profit inspectors believe professional landlords make.[18] Their minimal profits are one reason inspectors believe that small-time rentals are particularly vulnerable within the stacked deck.

A dog walker slowed to watch inspector Nick as he glanced from his notes to the building and back. We were standing, in the chilly autumn breeze, in front of a vacant three-story three-unit brick building on a wide, tree-lined residential street in a racially and economically diverse neighborhood. The building was in court for a dangerous garage and porch, and we were there to check on progress before its upcoming court date.

As we walked around the side of the building, Nick explained that the current owner—a middle-aged White woman from Eastern Europe—had recently inherited the building and she could not afford the necessary upkeep. Nick rubbed his furrowed brow and told me that no work had been done on the porch since the building's last court date, three months prior. Some masonry on the back of the building had also begun to crumble, and windows had been smashed since his last inspection. These were new violations, yet Nick did not add these to the existing list.

"I'm not going to add to her headache or ours," he told me. This was Nick's way of justifying his decision not to record these new violations. His compassion extended to building court the following week, where he gave testimony about the violations at the building. He could have asked the judge to give the owner an ultimatum: fix the issues or face fines and forfeiture. Instead, I watched as he recommended to the judge that the owner be given more time to remedy the issues, despite her lack of progress. I felt certain that Nick would not have shown such sympathy if it were a bigger building, or if he believed the building was professionally owned. He explained to me that he had done this because the owner was "struggling with money." The building's three units were all empty, so the owner wasn't making money from rent. Moreover, Nick told me that the location made the property a "developer's dream," and he thought that the building would be promptly bought, gutted, and rehabbed anyway. I did some research and Nick may well have been right. The neighborhood had seen steady development and upscaling in the past decade. Nick's knowledge of the housing market in the neighborhood and his sympathy for this struggling building owner aligned with his understanding of the stacked deck.

New regulations and contractor fraud are other factors in the housing market that inspectors believe disproportionately threaten small rentals. Inspector Steve told me that recent changes to porch regulations "hurt three-flats," "ma and pa buildings are what we call them," he added. "These are buildings that are probably owner-occupied and are not making lots of money for the owners. So, $35,000 is a big deal to them." John, a heavy-set Black inspector, described situations he faces on the job when owners of family buildings cannot afford to maintain their buildings because the odds are stacked against them.

> A lot of the building owners from the '80s and before . . . had the desire and they had the financial means to upkeep the building and then, you know, cost

of workers, cost of labor, all of that was, you know, the cost of materials was much cheaper then. Inflation has just made everything so expensive now. You have a lot of slumlords, and you have some that just don't have the money to do it.... Sometimes you have a landlord who wants to do it, he just don't have the means ... or you know maybe a family-owned building ... it might be a three-unit in which the owner may live on one floor and family live on other ones, so then we're a little bit more cognizant of that situation. To tell a person that you got old windows ... you need to replace these windows. To say that to someone who I know is on a fixed income, that has a family, [and] none of them are complaining about it, they understand the problem, they want to get it fixed, they're trying to go through programs, but they don't have the money.

The details John provided — and the heartfelt way in which he spoke — made me think he or his family, who live in a working-class neighborhood on Chicago's West Side, might have neighbors in this very situation. I asked him what inspectors like himself could do in these situations. "We'll help," he responded immediately, "because those violations that's not dangerous or hazardous, they're subjective. And there are certain things that I can do to help them as far as get permits and that stuff. . . ." He trailed off, perhaps concerned he had said too much, and changed the subject. "I know that they're doing the best that they can with as far as they're trying to get the money. And so even if maybe there's a year or two [before the violations get fixed]," he told me, without finishing his thought. John, and other inspectors as well, make stabs at justice — by overlooking some violations and not insisting they get fixed promptly — for property owners they perceive to be "doing the best they can" when the deck is stacked against them.

Inspectors' concern for small rental buildings is warranted, because they are disappearing. Recently, two- and three-flat rental buildings were the only housing type in decline in Chicago. In tight markets, building owners and developers gut the interiors to build single-family homes or condos. In poorer areas, these buildings are disproportionately left to deteriorate and eventually end up demolished.[19] In each context, the stock of small rental buildings is diminishing. Small rental buildings have long been a way for property owners to earn an income while providing much-needed rental units that tend to be affordable, meaning that their disappearance further stacks the deck against the city's renters. But, as we see later in this chapter, inspectors are not motivated by the plight of tenants when they go easy on small rental buildings.

In a chapter about rental housing, it makes sense to assume that the word "rent" refers to the amount landlords charge and tenants pay for housing. But the word has a broader meaning—denoting any profit derived from income produced by scarce assets.[20] Under this definition, landlords are rentiers because they make money from housing. Scholars of rentierism make it clear that profiting from an asset requires effort. It is not enough just to own an asset, such as a building. Rentiers also have to work to make their assets profitable. And this clarification sheds light on inspectors' adjudications of fair versus unfair profit. They deem landlords' profit to be fair if they see evidence of effort and hard work.[21] This evidence can take two forms. As we have seen, inspectors look for evidence of effort and hard work in the physical condition of buildings and assess conditions as a case of malign neglect in the absence of such evidence. But they also discern evidence about effort and fairness of profit from the position of landlords within the stacked deck.

In sum, inspectors do not frown upon all profit, or all rental properties. When inspectors assess two-, three-, and four-flat rental buildings, they are less likely to record violations.[22] This is not because small buildings are perfectly maintained. Nor is it because owners and landlords of small rental buildings eschew profit. Inspectors believe small-time landlords face financial struggles, and thus make stabs at justice by rarely recording minor violations and seldom insisting on compliance. Inspectors protect small-time landlords from additional expenses out of sympathy, not just because they assume the landlords cannot afford to make necessary repairs or to evade extra work themselves. These kinds of pragmatic considerations are part and parcel of inspectors' decision-making rather than alternative rationales. Inspectors use the tools at their disposal—such as the size of rental buildings—to categorize landlords and direct their stabs at justice. These decisions incorporate on-the-ground assessments of housing conditions and interpretations of the financial status of landlords, but also calculations about how the deck is stacked.

But who are inspectors really helping when they go easy on small-time landlords? After all, inspectors do not always meet landlords or tenants on inspections, so they do not know for sure that the two- and three-flat buildings they inspect are family buildings rather than the property of corporate landlords. More broadly, bias toward small-time landlords can threaten progressive rental market regulations. Few people openly defend "landlords" writ large, because the word carries negative connotations of negligence and ruthlessness. But when we talk about landlordism in terms of small-time, family buildings, this can make housing regulations seem like attacks on ordinary people just trying to get by.[23] Small-time landlordism is couched in the rhetoric of enterprise and small-scale entrepreneurialism, and hailed as offering the last bastion of affordable housing. But some research reveals that, on average, even small-time landlords are wealthy, debt-financed, and integrated into the corporate landlord landscape.[24] This research casts some doubt on whether inspectors' assessments are always accurate. Irrespective of accuracy, their categorizations reveal the importance of relative assessments in the city. We can find more evidence for this in the following pages, when we see inspectors opting not to protect small-time landlords when they are implicated in neighborhood crime.

The Building Police

Can replacing a toilet seat really prevent crime? I asked myself this question as I sat in building court one afternoon. The court inspector was reading a long list of code violations out loud for the judge. Court inspectors give testimony in these kinds of cases, in which property owners are summoned to court if their properties have serious code violations. But this list included a missing toilet seat—hardly a serious code violation! This violation was one of many that had been reported in the two-flat residence of a local rapper (and alleged gang member) and his grandfather. The family building was on a quiet street comprising mostly two- and three-flats near a small playground on the North Side of Chicago. Two weeks earlier the peace had been interrupted by a SWAT standoff lasting over two hours, initiated in response to reports of gunshots. A rare event in the neighborhood, the standoff prompted media attention and alarm among neighbors. It also provoked a Strategic Task Force (STF) inspection—a procedure in which a team of building inspectors comb through properties with alleged drug or gang activity under instructions to record as many violations as possible. The aim of such a sweep is to use building court to force building owners to evict tenants that the police and local elected officials deem problematic.

The STF inspection in this small rental property had turned up twenty-nine code violations, ranging from a lack of water and broken roof tiles to high weeds, peeling paint, and that missing toilet seat. The resident—whose alleged ties to gangs and guns had precipitated the inspection—was murdered before the case went to court. When the court date came around, however, eighteen residents of neighboring buildings appeared before the judge, alongside a representative from the alderman's office. The property was already in foreclosure, but these community members seemed intent on hurrying the process along and seeing the back of the current residents.

The representative from the alderman's office herded the group of neighbors to the front of the room, and a journalist scribbled in the background as the hearing began. Though the residents of the building in court were Black, all but one of these neighbors appeared to be White. The judge—a White woman seemingly in her fifties—thanked them all for attending as she adjusted her robe, assuring them that "the court is very receptive to community concerns and turnouts like this." She listened intently as the advocates complained about the lifestyle of the owner's grandson and the loud parties that occurred at the property. From other building court hearings I had observed, I was used to hearing inspectors weigh in and suggest the court grant small-time landlords more time to fix issues—just as Nick had done—so as not to add to a building owner's "headache." Not in this case, however. The court inspectors remained stone-faced throughout, speaking only to verify that the building was indeed in the condition noted on the report. It seemed clear that the inspectors had recast the owner of this family building as a bad landlord. Peering over her glasses, the judge promptly issued a vacate order on the property, declaring it unfit for habitation.

"We use the building code to fight crime," inspector Derek grinned. We were in the office, talking about STF inspections, which, Derek explained, are used as an alternative method of criminal justice:

> [The police] know they're engaging in illicit activity but sometimes, for whatever the case may be, they can't bring charges. In a lot of cases, those buildings were also in really bad shape. We were able to actually shut some buildings down or bring pressure on the owner to clean up his act by using the building code, which is I think pretty unique.

Bill shared Derek's enthusiasm, naming "rescuing buildings in distress" as a favorite aspect of his twenty-plus years on the job as we sped south on the expressway one morning just after rush hour. "By sending inspectors in to find violations," he continued, "[the City can] basically force owners to give buildings up." Bill was amenable to the City using inspectors like himself to tackle problems. "Sometimes you use the building code in your favor," Lou, a White inspector in his fifties, told me,

> to get an ultimate goal . . . you know if it's detrimental to the rest of the neighborhood, you make sure that you use the building code to the city's best interest and the adjacent property's best interest, you know, for the ultimate goal.

Inspectors justify STF sweeps, even in small-time rentals, as part of a broader logic, in which they come down hard on buildings that are "detrimental" to other buildings. This can entail citing property owners for accumulations of trash and debris in yards, weeds growing across property lines, or falling bricks, for instance. The effect of these issues on neighboring property owners motivates inspectors more than the issues themselves. In other words, accumulation of trash in the yard of a small-time rental is no big deal—until it begins to topple over into a neighbor's yard. During STF sweeps, the name of the game is ridding neighborhoods of crime and activities that may be a nuisance for other residents. Inspectors—like Micky—spoke with enthusiasm about the efficacy of the sweeps. Micky described a building used by dealers that was "so bad that women walking kids to school would walk extra blocks to avoid it." An STF inspection turned up a crack in the floor that inspectors used to justify the building's demolition. "And the neighborhood held a party during the wrecking!" he told me with pride. "It's only anecdotal, but I swear crime [in the area] has dropped since."[25]

I found myself wondering what happened to the rapper's grandfather's property after the STF court case. Records show that, soon after the judge's vacate order, the property went through foreclosure and was transferred to the mortgage company. An LLC bought the building, but, according to Buildings Department records, they have yet to fix any of the twenty-nine violations. It seems unlikely that this is due to a lack of funds. A bit of online digging revealed that the LLC owns a handful of other properties in the city, some of which are valued at over $1 million. Instead of offering protection to family buildings, STF inspections can put property and profit in the hands of the very corporate landlords inspectors usually rail against. This irony is not lost on inspectors. "We're killing these property owners," inspector Eddie told me, shaking his head, as we talked about STF inspections, "and it's hugely profitable." "Hugely profitable!" he repeated with venom. Although Eddie was concerned and angered about some consequences of STF, the overarching rationale among inspectors was that these sweeps offered protection for other property owners. Yet, as Eddie warns, this may be at the expense of small-time landlords, and can serve the interests of the kinds of profiteers inspectors usually disdain.

Inspectors' attitudes toward STF inspections show that categorizations of deservingness can change. Inspectors do not make stabs at justice on behalf of all family buildings or small-time landlords. Their categorizations are malleable and relative to other positions in the stacked deck. When small rental buildings are detrimental to other property owners, inspectors recast them as undeserving, even if that means allowing profiteers to benefit.

Tenants

This section demonstrates that inspectors show tenants no mercy in part because they have no recourse to protect them. Yet they still use relative assessments—stemming from their perception of the stacked deck—to make distinctions between renters, from undeserving newcomers and housing voucher recipients to tenants who inspectors believe deserve better. That inspectors make these distinctions despite their inability to act on them points to the ubiquity of the stacked deck as an organizing logic.

No Mercy, No Recourse

It was a normal Tuesday morning for inspector Malcolm. He blew on the coffee in his thermos as he leafed through the pile of 311s held together with a blue rubber band. He tapped on one with his finger, as if to select it, and started the car. We were going to a basement apartment whose tenant had called 311 about poor ventilation that was making her sick. The tenant—an Italian woman in her seventies—was still sick, she told us when we arrived, even though it had been months since she made the 311 call. Her health problems—labored breathing, headaches, and a persistent cough—had started a few years ago when she moved in. She had moved to the neighborhood to be closer to her son, whom she pointed out in family photographs on a small side table. She was convinced that whatever was causing the unit's dampness and musty smell was the culprit for her deteriorated health.

The landlord lived upstairs and had had a bad reaction (according to the tenant) when she heard about the call to the City. Yet she had done nothing to remedy the situation. Based on the lack of windows and low ceilings—and the frequency with which he had to duck to avoid hitting his head on the ceiling—Malcolm deduced that the dimly lit basement apartment in the three-flat was an illegal conversion, meaning the owner would have to go to court. I was surprised by the tenant's reaction to Malcolm's decision. She implored Malcolm to "show mercy," explaining that she just wanted to move out as soon as possible and feared her landlord would retaliate if she was brought to court. Retaliation could entail the landlord finding a way to keep a security deposit, writing a bad reference, or making her life even more uncomfortable before she could move out.[26] Malcolm did nothing to put the tenant at ease. In response to the tenant's pleas, he rather impatiently told her that she would probably have moved out by the court date anyway. I was struck by Malcolm's brevity and stoic tone, which conveyed to me that he cared little about the fate of this tenant.

It seems like everyone—from neighborhood associations and elected officials to homeowners—assumes that tenants are not invested in the neighborhoods in which they live.[27] Presumptions about the transient and detached nature of tenants have consequences, prompting widespread beliefs that the presence of tenants decreases a neighborhood's collective efficacy—the ability of residents to control the behavior of other residents and create an orderly environment. Tenants are associated with increased crime and disorder and lower property values.[28] But inspectors have compassion for people against whom the deck is stacked, and the deck is stacked against tenants in US cities like Chicago. They are most at risk of displacement as neighborhoods change and affordable housing diminishes, and there are few government provisions to protect them from eviction at the whim of their landlords. The stacked deck motivates inspectors in other situations. So why did Malcolm not make a stab at justice on behalf of this tenant? Why do inspectors fail to sympathize with tenants and instead overlook their precarious positions within the stacked deck?

One reason inspectors do not attempt stabs at justice on behalf of tenants is a lack of recourse. Inspectors have no real options that either clearly help or hinder tenants, because code enforcement is directed at property owners.[29] Their decisions about whether to record a violation or bring a property owner to court, for example, entail actions against the owner—not tenants. To be sure, inspectors could be more compassionate toward tenants, but they do not disdain tenants to the same extent they do some landlords. Still, inspectors apply a familiar logic, based on relative positions within the stacked deck, when they interact with Chicago's renters.

Hating on Hipsters

"This one's going to be messy," inspector Nick warned the building court attorney. "This neighborhood is gentrifying." The Northwest Side neighborhood he referred to was making local headlines because of the spate of evictions and displacement of Latinx tenants. Sure enough, the landlord of the three-story brick building in building court was in the process of evicting tenants so that he could gut rehab the building. The evictions, according to Nick, had prompted the tenants to call 311. While he admitted that the condition of the back porch was a legitimate request—and the reason the building was in building court—he also suggested that the other issues the tenants had mentioned, such as a broken window, holes in the guttering, and high weeds, were not justified. "It's only the hipsters, with the Bernie Sanders signs out front, [who] want something for nothing all the time, [that are complaining]," he assured the attorney, sighing as he looked back at the paperwork in front of him. Here Nick made an important assumption: that it was the newcomer "hipsters" who had called 311.

Tenants are more likely than homeowners to be the first wave of—and a recognizable sign of—gentrification. Inspector Antonio decried the presence of a new art space and the arrival of White hipsters in the Latinx neighborhood where he once lived. "Yeah, I'm pissed," he said simply.[30] Inspectors are motivated by a particular image of a tenant as one who takes advantage, and they use the word "hipsters" to invoke a broadly maligned target. But inspectors' disdain for tenants does not discriminate between existing or longtime residents and newcomers. Marco, a Latinx inspector with over fifteen years' experience on the job, told me how he reacts when he knows tenants make complaints to get their landlords in trouble. He told me about a rental building in a high-poverty neighborhood on the South Side which tenants had "sabotaged" by removing smoke and carbon monoxide detectors and then calling 311 to complain about their landlord's lack of safety provisions. Marco was suspicious of tenants and did little to sympathize with their precarity. Inspectors did not discriminate in their lack of sympathy for tenants, however. They expressed disdain toward White gentrifiers as well as low-income renters of color. Inspectors' attitudes are both unsurprising and surprising, given what we know from existing literature about broad disdain for tenants and racism among frontline workers. Although inspectors tend not to sympathize with tenants, there is racial uniformity in their attitudes.

"Shit Out of Luck" versus Section 8

Occasionally, when they inferred evidence of struggle, inspectors showed concern for tenants. These exceptions help to hone in on the underlying logic that guides inspectors' discretion. Danny expressed sympathy for an

African American woman whose rental unit he had just inspected. "Now she's shit out of luck," he exclaimed angrily, as he told me she was a single parent and was managing to hold down a couple of jobs. "She has no Section 8 voucher; she's paying market rate for a shit hole." He paused for a moment before continuing:

> What I'm running into is people that are stuck in between. They're trying to make it in a market sense. They're trying to work. They have no Section 8 voucher to assist with the rent, no utility assistance, and they're trying to make it. And essentially the numbers and percentages of people who are trying to do that are vastly higher that what the general public believes. They all think they're all on welfare and it's just not the case.

To Danny, this tenant is a prime example of effort and struggle.[31] She is "trying" to work, "trying" to "make it." Danny notes—twice—that she is not a housing voucher recipient.

Danny's statement about this tenant who is "shit out of luck" despite her efforts is even more revealing if we compare it to his approach to housing voucher recipients. "I consider Section 8 in one of two ways," Danny told me. He said he would "automatically write the shit out of" a bad landlord who was profiting from the voucher program. However, if he found a 311 call to be unwarranted, then he gets "pissed at the tenants complaining," because he thinks that voucher recipients "have it good." While Danny decried exploitation on the part of landlords, he also fell back on a trope about voucher or welfare recipients being free riders. This is a common trope in US society. In this view, housing voucher recipients are getting something for nothing. Of course, this is not true; voucher recipients pay 30 percent of their income for housing.[32]

But we can learn something from Danny's logic. Danny's approach to housing choice vouchers reveals that calculations and assumptions about unfair profit and benefits—whether on the part of landlords or tenants—motivate inspectors like himself. Unlike voucher holders, he clearly believed the tenant who was "shit out of luck" was *not* taking advantage of the system or getting something for nothing. Danny's logic illustrates something important about how inspectors see inequality. We might expect them to support subsidies and state interventions into the private rental market because of their contempt for exploitation in the rental market. But their sympathies do not extend to tenants with vouchers. Instead, ironically, they subscribe to the stereotypes about voucher tenants held by many of the landlords they disdain.

Tenants do not fit neatly into inspectors' category of profiteer (because they do not earn money from rent) or their category of precarity (because they do not invest in property in a way that is meaningful to inspectors). However, inspectors still use evaluative tools and calculate proportions to make sense of tenants' position in the stacked deck. This suggests that these tools are so pervasive in the world of these frontline agents that they use them even in situations where they do not necessarily apply.

Conclusion

The elevator inspectors in Colson Whitehead's fictional novel, *The Intuitionist*, are divided into two camps: empiricists, who use objective instruments and traditional methods to assess the condition of the elevators they inspect; and intuitionists, who rely on their instincts to "intuit" the state of elevators.[33] Building inspectors in Chicago do neither. Instead, they assess the condition of the buildings they inspect by piecing together information about the physical condition of the building as well as the social conditions of the people who occupy or own the building. With this information, they make stabs at justice.

All frontline agents with discretion have the power to make stabs at justice. That is, they can dish out red tape selectively and strategically and force some people to jump through hoops while giving others a pass. But frontline workers can also be lenient. They can overlook issues and give people a break. By explaining the motivations behind these decisions, this chapter recasts leniency and punishment as stabs at justice.

Thinking about frontline workers' decisions as stabs at justice begs questions about decisions in a broad set of contexts. How do police officers make stabs at justice? We know that the police use their punitive power unevenly, but this chapter suggests that studying how and when officers also opt for leniency might help clarify their motivations—and how their decisions might be changed. Can we conceptualize the decision of a DMV employee to overlook a missing piece of paperwork, or send a customer to the back of the line, as stabs at justice motivated by the employee's sense of injustice? A medical coder's choice to check one box over another could be a stab at justice if one box means a patient's insurance covers a procedure and the other does not. Do government officials who allocate housing vouchers make stabs at justice, and if so, how?

Taking an in-depth look at stabs at justice in the context of rental housing reveals the evaluative tools inspectors use to categorize people and property. They need to be able to figure out where to slot people in order to take stabs at justice and direct leniency or punishment in what they see as the right direction. Inspectors' evaluative tools consist of relative evaluations. In other words, their assessments of buildings are not just about one person or building or neighborhood, but about its relationship to another person, property, or place. The stabs at justice of other frontline workers are likely to also rest on relative assessments, because stacked decks are relative phenomena. Other frontline workers use comparative categorizations to make decisions, but these may be more noticeable in situations where they deal with the full spectrum of a stacked deck. While police officers deal with a range of issues, their categorizations might be clearer if they dealt with tax fraud *and* petty drug crime, for example. Or if a government employee distributed housing vouchers to low-income renters *and* mortgage interest deduction tax breaks to homeowners. This chapter suggests that we look more closely at how all kinds of people use relative assessments—based on calculations of relationships within stacked decks—to make decisions.

CHAPTER FOUR

Helping Out Homeowners

Changing Faces and Stubborn Realities

The rusty side gate creaked as inspector Malcolm opened it, letting himself through to the walkway of a one-story, single-family home. The house was set back from its tree-lined residential street in a middle-income, majority White neighborhood. We were responding to a 311 call that someone had made about a metal fence with sharp edges. The fence in question separated the small single-family house from the twenty-unit apartment building next door. "Yup, the edges are sharp," Malcolm remarked, lightly tapping a few points with his finger. This was all Malcolm needed to cite the property owners. But his evaluation did not stop there. Malcolm looked back and forth—with a furrowed brow—between the single-family home and the apartment building, holding the 311 slip that mentioned the fence in his hand.

A commuter train rattled along tracks nearby. Malcolm's look of puzzlement dissipated as he explained to me that the 311 operator must have confused the addresses of the complainant and the apartment building. He assumed that the owners of the single-family home had complained about the *adjacent* property, but that the operator had instead recorded the address of the single-family home on the request. Malcolm opted not to cite the single-family home for this reason, telling me that if the fence was removed, there would be no guarantee it would be replaced, meaning "suddenly [the single-family homeowners] could have thirty tenants on [their] yard grilling." Malcolm does not just scorn big rental properties, as we saw in chapter 3. He also wanted to protect the owners of the single-family home. And he had gone to some lengths to interpret the 311 request, and decided on a course of action—not issuing a violation—to do so. The truth is, Malcolm had no idea who was to blame for the sharp fence, nor who would be responsible for fixing it. But out of concern that the owner of the single-family home would face negative consequences, he decided not to take the matter further.

Malcolm's effort to protect this single-family building was a stab at justice. As we saw in chapter 3, stabs at justice comprise small and immediate acts of leniency or punishment that are motivated by perceptions of the stacked deck. Like his colleagues, Malcolm tries to help out owner-occupied properties when he can by protecting them from the costs and hassle of repairs. This has been the case for at least ten years, and across the city. I went back to my statistical analysis of building violation data, and saw distinct patterns: areas with greater numbers of owner-occupied properties receive fewer violations per 311 request than those with more rental properties.[1] This means that even if two neighborhoods get the same number of 311 requests about buildings, the one with more owner-occupied housing—from single-family homes to condos—would have fewer recorded building violations. In contrast to rental buildings, building inspectors go easy on owner-occupied properties, shielding them from the headache of recorded violations. Inspectors' decisions to protect property owners align with historical and sociological accounts that document the valorization of homeownership in US society.[2] For many, homeownership is the crowning point in achieving the American Dream—something that the media, cultural rhetoric, government policy, and even the tax system teach us to aspire to and protect.

But inspectors do not treat all owner-occupied properties equally. As this chapter demonstrates, inspectors show the most leniency to owner-occupied buildings whose owners they believe struggle with upkeep. The chapter also takes us to changing neighborhoods to see how these contexts—and the precarity in which inspectors believe they put property owners—can prompt compassion for newcomers as well as existing residents, or those we might call "old-timers." We see inspectors using stabs at justice to even out the benefits of public-private initiatives and to steer property owners away from precarity. Next, we see inspectors' disdain for greed changing and expanding their categorizations of victims and villains. The final section demonstrates that inspectors show compassion in communities of color, where histories of wealth extraction have deprived property owners of the resources needed to keep up with maintenance. But, because of obdurate disparities in housing conditions, inspectors' hands are tied, and they end up giving out more violations than in White neighborhoods.

Overall, this chapter uncovers how inspectors interpret, navigate, and shape the physical and social fabric of owner-occupied housing in the city, revealing a whole host of logics and motivations that cut across common distinctions between property owners, old-timers, newcomers, and racialized urban residents. Yet while some perceptions of the urban fabric are changeable, material realities are more stubborn. This chapter unpacks this tension, revealing how limited stabs at justice are at making a mark on the stacked deck.

Categorizing Homeowners

Inspectors regularly take stabs at justice to protect homeowners, but this is not the whole story. Inspectors slot homeowners into different categories and prioritize homeowners they believe to be struggling to keep on top of maintenance. Their compassion even leads them to interpret disrepair as defensible at times.

Protecting Homeowners

It was midmorning in a majority White and upper-middle-class neighborhood on Chicago's Northwest Side. It seemed obvious which house we were here to inspect. I could just about make out the flaking white paint on the house beneath the ivy, which seemed to be taking over and extended to the wooden fences on either side of the house, which sagged under the growth. The house stood out among its neighbors. The other houses on the street were small and old, a mixture of frame and masonry buildings with established plants in their yards. They all looked neat and well maintained—not a paint flake to be seen on their iron gates and railings. Inspector Eddie and I walked around the side of the two-story frame house, stepping carefully so as not to trip over ankle-high overgrowth that crept across the narrow walkway. Eddie rang the front doorbell and, as we waited, I noticed two broken panes in the transom. No one came to the door, so we could not inspect the inside of the building. Back in the car, I watched as Eddie scribbled "no entry"—to note that he had not accessed the property—on the inspection report balanced on his knee, using his elbow to steady it as he wrote. He did not list the sagging fence, broken windows, and overgrowth in the yard. This was typical for inspectors, Eddie explained, who will "just do a 'no-entry' and call it a day for single-family homes." Eddie and his colleagues typically do not record minor violations when they inspect single-family homes.

But not all owner-occupied homes are single-family buildings. Many people who own two- and three-flats, for example, live in the building, and this can make it hard for inspectors to decide how to treat these buildings, because they do not always know if the property is owner-occupied. As inspector Malcolm told me, often "there is no way of knowing" if a small multi-unit building is owner-occupied when inspectors turn up. So he tries to find out. I watched Malcolm counting electricity meters for this purpose. If there are three meters for a three-unit building, for example, this leads him to believe that one meter covers a unit plus the building's common areas, and thus that the owner lives in one of the units. Malcolm goes to noteworthy lengths to deduce the tenure of properties, considering how busy building inspectors are. He does so because this is such a significant tool of evaluation for inspectors. Malcolm and his colleagues want to know whether buildings are owner-occupied so they can make stabs at justice on behalf of owner-occupants.

Struggling Homeowners

"For the most part, your regular owner-occupied building, they try their best because nobody wants to live in a pigsty or deplorable conditions," Dave (a White inspector with twenty years' experience) told me during an interview. "We don't nitpick," "we're not out to hurt a homeowner just trying to do the right thing with his property." When inspectors talk about property owners "trying to do the right thing," they are referring to keeping on top of maintenance and fulfilling the responsibilities of homeownership. Other research suggests that maintaining a property—by mowing the lawn or keeping up with repairs over a number of years, for instance—can really make a difference to the perceptions of neighbors, city officials, and others. Debbie Becher, for example, demonstrates that city officials tend to value property through a lens of investment—in terms of time and energy as well as money—which has important implications for how property owners get treated by the government.[3] And Claire Herbert finds that this appreciation for effort even extends to squatters who maintain properties.[4]

Inspectors respond to signs of investment by showing property owners leniency. Inspector Antonio elaborated:

> We'll go to a property [and] if the property's clean, you know, then we'll walk away and put "no apparent violations" or "no cause." And we'll find a lot of

them like that, if the house is neat, it doesn't have to be brand new, but if it's neat or well-kept, we'll leave it alone.

In this instance, Antonio suggests that some material clues, such as cleanliness or being "neat or well-kept" are evidence of effort in inspectors' eyes. But, as in Becher's account of city officials, inspectors are particularly sympathetic to property owners whose effort and investment are—they believe—out of proportion to their resources.[5] Inspectors make assessments based on assumptions about precarity.

Economic segregation in Chicago means that it is not hard to guess the approximate socioeconomic status of a property owner from their block or neighborhood. Inspectors sympathize with property owners who they assume lack the resources for maintenance, and try to help them out by not recording minor violations. Inspectors' compassion toward low-income property owners is surprising, considering what we know about the tendency of frontline workers to punish the poor in other contexts. So, it's important to verify that what inspectors were telling me is really what is going on. I went back to the data set I had compiled of building violation data and 311 requests, matched with demographic survey data, to try to do just that. Sure enough, what inspectors were telling me bears out in the numbers: my statistical analysis of the data set reveals that there are fewer recorded building violations in areas with low median incomes than in areas with higher median incomes. And this pattern holds even when I account for the number of 311 requests that come in for each area. It also holds when I account for the average age of buildings in the area, which is a factor in building violations—as buildings age, they need repairs. So, it seems that—just as they attest—when inspectors turn up at homes in low-income communities, they are less likely to cite buildings for code violations than they are to cite buildings in areas with higher incomes.[6]

Defensible Disrepair

Maintenance and effort are signs of investment for inspectors. But inspectors also sort some properties with a *lack* of maintenance into the meritorious category of the struggling homeowner. Over coffee one afternoon, inspector Danny recounted an inspection in a Latinx neighborhood that was still on his mind from that morning. The neighborhood had been in the news recently for alarming asthma rates, polluted air, and the nearby coal plant. "I'm not gonna write them up," he told me, before pausing to sip his drink.

It's just a single-family home. Sure, the gutters are hanging off, [they] got a couple broken windows, [but] the heat works, the hot water works, they're dirt poor. I'm not gonna write them up. No one's gonna die there.

Inspector Derek echoed Danny's sentiment: "If somebody didn't have money to fix their gutter or whatever the case might be, I don't want to write them a violation to add to their misery." In cases like these when a homeowner is "dirt poor" or "didn't have the money," inspectors categorize violations as *defensible disrepair*—a term I use to capture the excusable and understandable nature of some code violations.[7] Danny made sure to reassure me that he would not turn a blind eye if conditions were dangerous. But in general, inspectors think it is unreasonable to expect property owners in modest buildings in low- or moderate-income neighborhoods to keep up with maintenance. Their assessments of defensible disrepair are based on their assumptions about how the material conditions came to be—namely, the effort and socioeconomic status of the building owner. I listened as inspector Danny—a White and working-class inspector of fourteen years with a loud booming voice and a beard—recalled trying to help an African American woman whose "family building" was in such a bad state that it was close to being recommended for demolition. He told me that the woman was the single mother of seven children and had been holding down numerous jobs until her car broke down, causing her to lose her jobs. "What I'm running into is people that are stuck in between," Danny stated:

> they're trying to make it in a market sense . . . but their mortgage got out of control, they're in foreclosure, they failed to tap into the HARP program to get their mortgage modified, and their principal, interest, and taxes are $3,000 a month. And that's just outrageous for a two-flat on the West Side. They could have used a property tax appeal and would have won, [which would have] reduced their property taxes.

Inspectors like Danny are savvy about federal and local property law and policies. Danny's sympathy for this woman was rooted in her financial status, but also in his outrage at the system of property tax assessments and appeals and the inefficiency of the federal Home Affordable Refinance Program (which was established in response to the housing crisis to help struggling homeowners refinance their loans). There are rules for what makes a building "demo-worthy," in the words of inspectors, but the final recommendation requires inspectors to score the building on a scale of 1–100. Danny was able to cite the building for its severe issues but not recommend it for demolition. His benevolence stemmed from the owner's struggle to "make it" as a property owner, and the effort involved therein, when the odds were stacked against her.

The category of the struggling homeowner thus includes property owners who inspectors believe make an effort out of proportion to their resources, as well as owners who, in inspectors' eyes, cannot reasonably be expected to maintain their property. The underlying thread is inspectors' compassion for low- and moderate-income property owners. This is in sharp contrast to how they treat homeowners in wealthy neighborhoods, or who they suspect are willfully negligent. As inspector Nick told me one morning when we were looking at a fence that had been partially knocked down in front of an inconspicuous two-flat: "It's not [due to] to neglect, so why hammer things?" His statement suggests that if the issue *were* due to neglect, then he would come down hard on the property owner—or, in his words, "hammer things." Inspector Paul elaborated on the distinction inspectors make between owners with and without resources, telling me that "if someone actually has money . . . then I'll write them up for the small violations." A reason for Paul's disdain is that, as with big professionally managed rental properties, inspectors assess violations in wealthy neighborhoods as malign neglect; they believe there is no reasonable excuse for code violations in properties owned by people with wealth and resources.

Neighborhood Change

This section reveals that the distinctions inspectors make between homeowners can change as neighborhoods do. We see inspectors make both familiar and novel decisions about who counts as a virtuous old-timer and a welcome newcomer. We also find out that inspectors make careful delineations between good and bad investment and neighborhood development.

Helping Out Old-Timers

Once colorful flowers were dying in the yard next to two faded plastic chairs. It was the middle of a long day of inspections, and inspector Malcolm and I were at a two-flat brick building that had a basement conversion, but no City permits on file. We were in the northernmost part of a historically Latinx area of the city. Whites had just come to outnumber Latinx residents in the neighborhood—which Forbes had recently rated as one of the twenty "coolest places to live" in the US. This coincided with the construction of new housing developments along a segment of a disused elevated railway line reinvented as the "606"—a 2.7-mile landscaped trail for pedestrians, runners, and cyclists. This kind of neighborhood change exacerbates the precarity of old-timer residents, in the eyes of inspectors.

Inspector Malcolm suspected that development companies were responsible for making 311 requests in changing neighborhoods like this, in an effort to drive down property prices or prompt owners to sell in lucrative areas. He told me that, contrary to "an average homebuyer who likely wouldn't know what was an illegal basement or not," developers are savvy about "what violations are and what counts." Malcolm's suspicion fits into a broader belief among inspectors about the high volume of 311 calls in neighborhoods as new people move in. "You get people complaining because they don't like the neighbor that they bought a condo next to," inspector Matt told me, "but [the building's] been that way for twenty to thirty years." Neighborhood old-timers deserve leniency, in inspectors' eyes, in part because they are targeted by others.

Two children cycled past on squeaky bikes as we pulled up at the building with the faded chairs and dying flowers. We were met by the owner, a Latinx man with a heavy accent who appeared to be in his sixties and who wore a faded baseball cap and jeans. The owner showed us into the now empty basement and pointed out that the kitchen that had been in the corner had been removed, meaning he could no longer rent it out as an apartment. "You know you can't rent this out, right?" Malcom asked a couple of times, "no one can live down here, ok?" The owner's nods seemed to persuade Malcolm. "This will be dismissed from court then, if you agree not to use it as an apartment." The owner nodded again, this time with a nervous smile, before cautiously asking Malcolm's opinion about replacing the building's garage. Malcolm glanced toward the building. "It looks to be in decent shape," he responded, "and you'll have to get permits. I can give you information on that, but it looks decent from what I can see."

On top of offering to assist this property owner, Malcolm opted not to follow up on the retroactive permit for the basement—nor to write him up for some issues with the basement windows that he had noticed. Each of these decisions was a stab at justice. As became customary, he explained his decision to me as we got back into his truck. "Those window issues are only small," he told me, so "I won't cite them." "What about the deconversion permit?" I asked. Malcolm stopped with his hand on the ignition key. "This guy, in this two-flat, probably doesn't have a lot of money," he explained, "plus he's just lost the revenue from renting out the basement." Malcolm estimated that the owner was losing $600–700 per month from the rent on the basement. "I don't want to pile on extra for this guy," he stated, before elaborating further.

> This [unpermitted conversion] will remain on his building's file and can always potentially come up in the future, especially now the public can see violations online. It could also hold up loans from the bank. But, making someone get a permit means they have to pay an architect $2,500–$3,000 for drawings and they are basically just a drawing of the existing basement without a kitchen.

Malcolm didn't make this decision lightly. He had reservations, but he wanted to do what he could in this moment to help the property owner. Malcolm, who grew up in a working-class neighborhood himself, assumed the owner did not have much money and was an old-timer in this rapidly changing area. We'll find out about the implications of Malcolm's decision—and whether his stab at justice was successful—in chapter 5.

"Good" Gentrification

"Now I'm going to show you an example of a good urban development project and gentrification," inspector Nick said, as he looked up from his pile of 311 requests and started the car. His use of the word "gentrification" conjured specific images in my mind: loft conversions, coffee shops selling pour-overs, and apartments featuring exposed brick occupied by young White renters. But this is not quite what Nick had in mind. He was talking about an existing resident converting some empty office space into residential units. We drove through the remnants of rush-hour traffic to the two-story brick building, which was on the North Side of the city, a few blocks from the lake. The building sat on a side street opposite a parking lot for a local school, almost beneath the El tracks, and a block from a shiny new Whole Foods and multiple newly rehabbed apartment buildings. The building looked plain and unremarkable, the kind of place you might walk by every day without noticing it. There was no exposed brick in sight. As he swung the car into a parking space, Nick told me that the owner had converted part of the building into residential units after it had sat empty for years. Nick paused for a moment when I asked about the cause of our court-mandated inspection. "Well," he said sheepishly, "unfortunately he did it illegally."

A middle-aged Italian American man named Roberto owned the building. He had converted the space without permits and was currently in the process of "trying to make it legal"—as Nick put it—by hiring an architect to draw up plans and applying for retroactive permits. As we were leaving, Roberto shook his head. "The City of Chicago [is] tough on people," he said when Nick was out of earshot, "they want money. I would like to tell them: 'Listen, we're not made of money. We work for a living.'" But Nick is "one of the good guys," he added, "he knows what people go through." Roberto was referring to Nick's experience in the city. Roberto also confided in me that Nick had given him "a few suggestions" on what to prioritize. Nick had been patient with Roberto, advising him to concentrate on the permits before fixing some of the other violations on the building's record. And when Roberto was late showing up to building court the following week, Nick assured the city attorney—and a representative from the bank that had funded Roberto's renovation—that everything was in order. To be sure, Roberto could have been planning on renting his new units at high prices, or providing affordable units in an increasingly expensive neighborhood. Nick did not know. But he opted to help Roberto out. Nick took a stab at justice by attempting to steer potential rental income to Roberto, a current resident.

Nick, Malcolm, and their colleagues are motivated by something that Japonica Brown-Saracino calls virtuous marginality.[8] For inspectors, the virtuous marginality of the residents they meet is often determined by the relational qualities of the stacked deck. Old-timers are worthy when inspectors believe they are most at risk from their neighborhoods changing around them, and from the deck being stacked further against them.

Inspectors weigh up the potential for new neighborhood development to benefit old-timers. "I'm taking pictures," inspector Javier began, enacting taking a photograph, holding his hands to his face and pressing an imaginary shutter-release button. He was recounting a case in which he discovered that a woman was living in a building that had an entirely collapsed back wall. Hers was the last single-family building still standing on a short block by the lake. New multi-unit developments had recently been built all around it, but she did not want to sell.

She looks out of the window and starts cursing me—"what you doing taking pictures? I live here." And I'm like, "you can't live there." So, I had to call supervisors, had to get the police out there, and we had to put it in court right away. And we actually got her to vacate and got her out of the building. She was an older woman, so this was probably where she was raised, that's what she was attached to.

Finally, we got inside, and it was deplorable. She had no plumbing, she was using buckets as bathrooms and actually at the end of the day it was a good story because they found her housing, she's actually living with her daughter. She sold the property to a developer, she got about half a million dollars. At the end of the day, after court, she actually hugged me crying. So, it felt good, you know, it shows that we have a purpose out there.

Whether or not Javier approved of the development that would replace the woman's family home, he seemed glad that this old-timer was able to benefit in the process.

"This Is Racial Profiling"

As I was driving through the city with inspector Eddie one morning, it struck me that inspectors sometimes extend the category of virtuous marginality to entire neighborhoods. This happens when inspectors believe residents are unfairly targeted. It was just after 10 a.m., and Eddie and I were driving through street after street of small bungalows on the Southwest Side of the city. Every street looked the same to me. But these modest homes are remarkable for being a bastion of post-war racial segregation. Their residents had been successful in their campaign to intimidate Black war veterans looking to move in, as well as in their opposition to public housing.[9] At least two building inspectors I knew grew up right around here, I realized. These days, the neighborhood is predominantly Latinx.

As had become my custom when we drove around the city, I asked Eddie what kinds of issues cropped up in the area we were currently in. I had barely finished asking the question before he started talking. He had noticed a spate of requests about illegal conversions recently, meaning that homeowners were using attics and basements as living spaces. This was a prohibited practice without first obtaining permits (to ensure that living spaces get enough light and ventilation and meet fire safety standards). Eddie continued to tell me that many of these bungalows did have illegal conversions, meaning an inspection could force owners to de-convert or get retroactive permits. Either of these options is a costly endeavor that could encourage—or force—owners to sell up or lose possession of their homes.

But Eddie wanted me to know something else: that a previous wave of German American immigrants had been the ones to illegally convert these buildings in the mid-twentieth century, and the City had turned a blind eye. "Now suddenly they're getting inspected and taken to court. This is racial profiling!" he told me, as we pulled up to a gas station. I couldn't see if the car needed gas or if Eddie just wanted to stop the car to concentrate on talking. He continued to tell me that White neighbors are more likely to call 311 on their Latinx neighbors. But he also thought perhaps someone with a vested interest—developers or someone in local city government—was lodging these requests to encourage dispossession and to clear the area for potential new development.[10] Whether or not his suspicions were correct, his assertion reveals the importance of the relationship between profit and precarity in his assessment. In this case, Eddie's beliefs about the profit-driven motivation of 311 calls cast the Latinx homeowners as particularly precarious.

But precarity can be hard to identify, and it can take unexpected forms. In changing neighborhoods, for example, it is often less clear for whom, and against whom, the deck is stacked. We often use categories of old-timer and newcomer when we discuss neighborhood change. But when we think about it, not all newcomers signal gentrification, not all long-time residents are virtuous, and not all precariously housed residents are old-timers.[11] This is one reason inspectors often find themselves debating whether to protect old-timers or newcomers, flitting between these categories using assessments of the neighborhood and scales of investment.

Scales of Investment: From Urban Pioneer to Family Man

This man is the epitome of a first-wave gentrifier, I thought to myself, as I caught sight of the White twenty-something who wore his flannel shirt tucked into his ripped designer jeans. First-wave gentrifiers are middle-class newcomers who clear the "frontier" of existing populations and pave the way for other waves of newcomers. Paving the way, gentrification scholars theorize, entails the restoration of buildings in disrepair, a process by which newcomers can assert a cultural claim to a neighborhood on top of an economic one.[12] The flannel-clad man had recently bought a two-unit building on a run-down street in a predominantly Latinx West Side neighborhood. The building's brand-new porch and siding meant it stood out among its dilapidated neighbors, some of which appeared vacant, with boarded-up windows, unkempt yards, and crumbling concrete stoops.

Inspector Nick was here to determine whether the rehabbed building—which had until recently housed around twenty day-laborers in a filthy and cramped basement—was fit for habitation. As Nick slowed to park the car, however, he told me he already knew that "the answer will be a no" and pointed to the front stoop and its missing handrail. "This is not ready to be inhabited." The discovery of the missing handrail seemed to set the tone for the inspection. Nick was unfriendly to the owner. His mood only worsened upon seeing the work that had been done beyond the scope of the permit, including a new kitchen and stairwell.

Suddenly, however, Nick's demeanor changed. The owner casually mentioned that he and his wife—who was expecting their first child—planned to live in one of the units. Nick made no further mention of needing a new permit or the building not being ready. Instead, he asked the owner to email a picture of the handrail once it was in place. Nick's sudden change of attitude was prompted by what he had learned—that a young family would be living in this building and the owner was not planning to be an absentee landlord or to flip the property and move on. "He probably went a little overboard on what his permit allowed, but he'll be living there," he explained. "It's not a great area, but it's a quiet block," Nick told me as we got back in the truck, "and we want properties occupied." He had opted to smooth things over for this new owner.

While in another context the owner may have represented a gentrifier or a profiteer, his narrative convinced Nick that he was also a deserving homeowner who would raise a family in the area. This combination of investment and heteronormative family structure made this newcomer—whom I had identified as a pioneer—as virtuous in Nick's eyes as an old-timer. Contexts of changing neighborhoods shape who is on the receiving end of inspectors' benevolence. Inspectors protect old-timers against the threat of gentrification, and sometimes help old-timers benefit from gentrification. Yet they will also protect newcomers—if they are homeowners, if their family structure fits the bill, if they occupy previously vacant property, and if rehabs are modest.

Inspector Danny is on the same page as Nick. When possible, he told me, he uses his discretion to "stimulate the market" in "disinvested areas" because he "give[s] a shit about the neighborhood." I asked him what he meant by "disinvestment." "Economically depressed," he responded, "after White flight." Clearly, to Danny disinvested neighborhoods are neighborhoods of color. "Giving a shit about the neighborhood" means Danny will not cite new residential or commercial developments for minor violations—things like "cracked windows and a few crumbling bricks"—or what he called "just housekeeping." He overlooks issues like this, he told me, to avoid penalizing property owners "in a way that discourages development." Building inspectors respond to the contexts with which they are faced. And the precarity in which inspectors believe neighborhood change puts property owners can prompt them to make stabs at justice on behalf of newcomers and new investment as well as old-timers.

Profit from Precarity

Additional political and social contexts can change inspectors' categorizations of villains and victims. Inspectors' wariness about undeserved profit and greed puts them at odds with city programs for home repairs, but they also try to steer some homeowners to use this program to their advantage. And inspectors' continued disdain for greed on the part of multinational banks, insurance companies, and fraudulent contractors sometimes causes them to overestimate the pervasiveness of the stacked deck.

Steering City Programs

Inspector Eddie apologized at least three times for being late as we arrived at a nearly vacant two-story condo building. The condo association—the group of condo owners that manage the finances and building maintenance on behalf of the other owners—was no longer functioning. This can happen when there are no owners willing to be part of the association, or when an incident or disagreement has caused the association to dissolve. The red brick building looked old, probably dating back to the early twentieth century, when the neighborhood's population was 99 percent White. Today, the population is almost exclusively Black. I expect that, as a group of White people huddled in front of this building, we stood out. Eddie and I had met up with two potential receivers, who were there to see if they were interested in investing in the property. Receivers invest in properties by stepping in to do court-ordered repair work in buildings whose owners are unable or unwilling to do so themselves. Once the receiver completes the work, they record a lien (a financial claim on the property title) against the property to cover the costs. If an owner does not pay off a receiver lien, the receiver can foreclose and take possession of the building.

A man named Heavy B let us into the building, where he lived rent-free and acted, in his words, as a "house sitter," which involved maintenance and keeping [the building] from "getting stripped" of copper piping and other metals. Heavy B made a good house sitter; the windows and brickwork were intact and the grass in the yard was freshly cut. And, for the year and a half since he moved in, he told us proudly, he had had no issues with crime—the building had not been touched. "Good job, Heavy B!" Eddie said, shaking his hand and grinning. The exterior of the building was in decent shape, but signs of water damage plagued the interior, and a moldy mattress sat under the first-floor staircase. The air inside felt cold and stale despite the hot summer day.

Turning to climb the staircase, I glimpsed the building's only other occupant: an elderly, frail-looking Black woman, whose coughing reverberated around the practically vacant building. Eddie's pace meant the receiving agents were almost running behind him, pointing out issues with the building—a missing hatch, patches of exterior brickwork that had not recently been tuck-pointed, and a PVC tube lodged in the cement basement wall. Eddie had either not noticed or did not care about these issues. Instead, he was noting the missing furnaces and exposed wiring in some of the units. Though the issues the receivers mentioned were less severe, they were all building code violations. Perhaps they were trying to help Eddie by pointing out these issues. Or perhaps they were trying to increase the number of violations on record for the building, to drive down the sale price or drive up the repair costs. A lower sale price or a higher repair bill could both increase profits for a receiver. Whatever the reason for it, the misalignment between Eddie and these receiver agents was stark. Eddie seemed unperturbed, though. Back in the car, he had still only written down the issues with the wiring and furnaces. Eddie's motivation for doing this was unclear to me in the moment, but I came to see it as another stab at justice.

Receivers invest in buildings when owners are unable to maintain their property. This sounds like a way to even out the stacked deck, and thus like something with which inspectors would be on board. But inspectors do not see receivers as a positive force in the city. Instead, inspectors view receivership as a way for someone to profit from someone else's misfortune. "They're like vultures," Eddie remarked as he almost ran a red light in downtown traffic one morning. Eddie was telling me about the receiver agents who sit in on building court cases, waiting to approach building owners when judges recommend receivership. I saw them too. While most worked for the city's largest public-private initiative, some were individuals who—Eddie alleged—get tipped off about opportunities to make large profits by their attorney friends. Inspectors were particularly critical of private receivers, assuming they are well-connected people getting tip-offs about lucrative investments. Their cynicism also extended to the public-private receiver agencies. "It's all cronyism," Danny told me nonchalantly over coffee, "it's all political payback." I asked him to elaborate.

> [Receivers are] nonprofit, but obviously that's a relative term. The administrators of a not-for-profit can make a nice salary. And that's saying nothing about exorbitant fees that can be channeled someplace that we know nothing about. Also, who do they market and sell properties to? You can steer it any way you want to. And trust me, there are some choice properties that I've seen marketed and sold.

Inspectors' contempt for receivership stems from the unfair profit—resulting from cronyism and connections rather than hard work or struggle—that inspectors believe is often made by some from the precarity of others. Inspectors do not see investment in property as universally positive. They value economic and emotional investments in a property, when made by property owners.[13] But they are leery of external investments that they suspect further stack the deck against struggling property owners.

Receivership also offers inspectors a way to intervene and take stabs at justice on behalf of homeowners. I listened as Danny attempted to steer the receivership process so that a condo owner could control and perhaps even profit from the investments in his building. The building in question, which looked more like a run-down motel than what I think of when I imagine a condo building, was on the verge of being taken over by a receiver to pay for some costly repairs. Inspector Danny and I pulled up to the two-story complex on a street on the edge of an ethnically and economically diverse neighborhood on the city's far North Side. The neighborhood was known for its diversity and culture, but this building felt like we were far from the neighborhood's bustling main street of restaurants, shops, and murals. The building was set back from a busy road, but traffic was constantly audible. Faded plastic children's toys were strewn across the yard, and two windows on the second floor were broken.

"My general impression?" Danny asked unsolicited, as he searched for a pencil in the pockets of his khakis. "A bit of a crap hole." The building needed a new roof and porch, and Danny cautioned me as we tested the wobbling wooden exterior stairs, shaking his head and sighing as he told me which steps to avoid. A unit owner appeared from around the side of the building and introduced himself as a member of the condo association board. He was around sixty and spoke with an Indian accent. He happened to be on the condo board for the building. He fully understood the issues with the building, he claimed, and planned to hire a contractor to take care of tuck-pointing the masonry. But the association was near bankrupt and could not afford the tuck-pointing. Danny suggested that the condo board member act as a private receiver—like the ones for which he had expressed serious disdain—for the building. "Someone has to negotiate with the insurance company," he told him. "They will do everything in their power to fuck people." Otherwise, he continued "you might end up doing more work than you need to do." The board member's furrowed brow suggested he was not keen, but he promised he would "run some numbers" and consider Danny's suggestion. Danny opposed receivership when other parties profited but was keen to leverage the program to help the current owners.

Greed, Crisis, and Thieving Assholes

"It was a shit show," inspector Frank lamented as he began to explain how the unfettered profiteering that drove the 2008 housing crisis had created a whole new category of precarious homeowner. He was unable to shake his anger as he recounted an inspection of a five-story brick condo building on the lakefront south of downtown that had taken him the better part of a day.

> I found no heat, no hot water in the entire 60-unit building. . . . The elevator was out. Gang graffiti in the hallways. The association was corrupt, according to one of the owner-occupied unit owners. Major embezzlement of collected monthly association fees. $400 per month. Roughly 25 of the 60 units paying that important maintenance fee. Bottom line? Embezzlement and nonpayment equaled no gas bill, elevator, electric [in the] common areas . . . massive, multiple violations. . . . In short, a shit show.

Since the housing crash, inspectors find themselves dealing with the aftermath of condo fraud more than ever before. Corrupt condo associations embezzled their financial reserves, leaving no money to pay for utilities, let alone maintenance or repairs. Frank continued:

> Whether it's an irresponsible or criminally liable landlord or association, it's always the same. Rent and fee collection doesn't go into maintenance. The property declines ... until we show up. Dude, they're horrible places, and it's the kind of stuff I love doing. I've always looked at these situations as crimes, and I'm the building police.

Frank expressed delight in making stabs at justice by coming down hard in cases like this. To him, the building he had inspected was emblematic of a broader trend of corruption and negligence. The housing crisis exposed a plethora of housing market actors and introduced new profiteers onto the scene. Profit had clearly met precarity.

Profit also meets precarity, in the eyes of inspectors, in low-income neighborhoods. As we sat in his car between stops one humid afternoon, inspector Danny began pointing out the numerous worn-out roofs he could see on the frame buildings in the area. Many of the roofs on these buildings, as well as the siding, had not been maintained and were beginning to slide off. Danny, however, blamed contractors as much as owners for this. This is notable because, like many of his colleagues, Danny used to work in construction. "Fly-by-night contractors," he told me, are "thieving assholes," and often just nail new siding into rotting wood, especially in low-income neighborhoods, which he described as "target-rich environments for fraud." As Danny explained,

> I'm constantly running into mostly elderly people who are in owner-occupied buildings and they didn't have the money to fix anything. Concurrent with that, ten years ago the situation was a banking industry that was loaning everybody money. So, it was remarkably easy for these people to get second mortgages. Unfortunately, that turned into a situation where a percentage of these people were being defrauded. Contractors would have them sign documents that would get the second mortgage for them. So, they had the social security number, everything. ... The bank loan went through and they were bringing their paperwork back and these old folks were just signing away the whole building. The contractor wouldn't do the work, he would foreclose on the contracts and take them and their building away. That happened a lot.

HELPING OUT HOMEOWNERS   117

Danny was clear that contractor fraud disproportionately and strategically occurred in buildings whose owners were poor. While data on this topic are hard to come by, Danny is probably not wrong that contractors take advantage of low-income homeowners in order to turn a profit.[14] But contractors are not the only entity that inspectors lament, as inspector Des will tell you.

> Insurance companies? Biggest scam on the face of the planet. They prey on people's fears. You know what? I've been driving for thirty-five years, forty years, whatever, never made a claim, I've been paying car insurance the whole time, just putting money in their pockets. It's terrible. And I've never made a homeowner's claim on my house.
>
> And I don't think insurance companies should be allowed to revoke insurance because these people have obviously paid their premiums, never missed ... and now, all of a sudden, they don't have fire insurance. And the house goes up, catches fire and [they're] not insured. How is that possibly fair?

Like other inspectors, Des mentioned unfair profit and pointed his finger at a clear villain. In this case, it was the insurance companies who generate revenue from vacant properties by hiking up rates for neighboring properties. Here, Des linked the situation to his personal expenditures and feelings of helplessness amid "the biggest scam on the face of the planet." To be sure, Des might not have such strong feelings about insurance companies if not for his working-class background.

Vacant properties and foreclosures weighed particularly heavily on the minds of many inspectors I spoke to. Inspectors also believe unfair profit is the cause of the foreclosed and vacant properties they inspect. Frank brought up a foreclosed property he had inspected that was close to being recommended for demolition.

> It was a foreclosure.... God only knows what they're paying for homeowners' insurance. I'm sure that area's been red-flagged. And then the fucking mortgage ... guaranteed they're ripping them off on the mortgage. Because a lot of these banks, smaller banks or the mortgage companies ... I can't be sure, but I'm almost certain that there's a federal mandate [that] compels them to consider mortgage modifications using various unsavory parameters.

Frank lamented the host of housing market actors and institutions that unfairly profit from—and cause—the precarity of this property owner. Des, a White inspector who moved slowly due to an on-site injury when he was a carpenter, expressed comparable feelings of injustice as he recounted how buildings make it onto the city's demolition schedule.

> A lot of times we have foreclosures in the nicer neighborhoods. They just, mom or dad lost their job or whatever, they just lost the building to the bank and the bank isn't in Illinois, the bank's probably on the East Coast or something and, you know, they have a servicer watching the building, so and then all of a sudden it's open and then we find it. And now we have our case.
>
> I guess the banks, they look at it as, you know, how much do they have with the mortgage, how much is it going to cost to rehab it? Some of these buildings were purchased for . . . you know, from 2008 on everything was a mess, an absolute mess, and now the banks are reaping the benefits of that . . . that's why there are so many short sales, they see what they have tied up and they see what it's going to cost, you know, they don't want anything to do with it . . . bank sale, foreclosures, then demolition. The bank will perfect the foreclosure but not record the deed back in their name because they don't have to. So if we tear the building down, the guy that gets foreclosed on, not only is he foreclosed on, but he [has to pay for] the demo. I don't know what the laws on are on this, but I'm hearing this and like "how is that possible? How is that possible?"

Des expressed outrage about two things: first, that national banks were "reaping the benefits" of the foreclosure crisis; and second, that the homeowners down on their luck face not only foreclosure, but also responsibility for the costs of demolition. But these two points are connected in Des's mind, as relative aspects of the stacked deck. Profit is particularly problematic, in inspectors' eyes, when they believe it is made from precarity. However, inspectors can get carried away, for example, by overestimating the situations in which banks can profit from precarity.

Inspector Derek had seen one too many foreclosed, empty homes near to where he grew up. He told me what he thought was the cause of the issue.

> The owner was gone, bank foreclosed, but these banks—being in the business of making money on the South Side of Chicago—they wouldn't take it to title. . . . They call them. . . . It would be in limbo. There's a word for it, they call them zombie properties. I don't really like that word but that's what they were. You had no owner. The bank said, "well, what do you mean? We don't own it."

Derek explained that banks had refused to step up and take ownership of buildings—this is the definition of "taking them to title"—even after foreclosures had been filed. To inspectors this was not just negligent, as it left no clear entity responsible for tax payments, property maintenance, and preventing disrepair. In Derek's eyes, banks were even more villainous because of the profits they reaped from their denial of culpability. Derek believed there were more zombie properties in areas on the South Side of Chicago because there was profit to be made. My research into the matter suggests that banks do *not* make profit from zombie properties; in fact, they risk losing money. However, Derek's insistence that banks profit illuminates how he made sense of the foreclosure industry and who wins and loses. He overemphasized the exploitative role of banks. The housing crisis was, for Derek, such a clear example of greed meeting precarity that he piled on an extra example and overestimated the profit that banks made and the extent of their reach. Derek saw exploitation where it did not actually exist.

Race and Housing Conditions

In this section, we hear from inspectors as they endorse different explanations for dilapidation and disrepair in communities of color and overlook how Whites profit from the stacked deck. We see inspectors' compassion for struggling property owners in minority neighborhoods, but learn that stabs at justice are blunted by the obstinacy of material conditions.

Structural and Cultural Frames

Just like everyone else, inspectors use frames to make sense of what they see. Framing is a way of comprehending the world, which provides us with explanations for the events and situations we face.[15] One common frame for explaining dilapidation and disrepair is the "culture of poverty," which blames the individual for their marginalized social position and, in doing so, obscures the role of structural factors.[16] When I first starting spending time with inspectors, I expected them to use this frame, and so I was surprised to hear them reject it. "Believe it or not," Marco—a Latinx inspector—told me, "the family homes on the South Side, they try the hardest. And a lot of times they comply [with the building code] better than somebody that's got the money on the North Side." Here, Marco recognized, called attention to, and pushed back against ideas that low-income communities of color are to blame for their poverty. Marco's brief statement speaks volumes about how inspectors view the racialized landscape of the stacked deck. Perhaps this statement is not surprising coming from a Latinx inspector, who we may assume is attuned to Chicago's racial and ethnic inequality. But inspectors across the board recognize that the deck is stacked against communities of color. And this creates an unusual dynamic, in which a group of mostly working-class White men are motivated to show leniency in Chicago's Black and Latinx neighborhoods.

For example, inspector Danny—who is White and in his forties—framed the material conditions of many Black neighborhoods in terms of disinvestment. "All the industry left," Danny began to tell me about a once-flourishing African American neighborhood on the city's Near South Side.

> And to expect the Black community to continue to thrive when the businesses were all pulling out? The buildings remain, but they pulled out and moved to the suburbs to a steel shed–like building with twice the square footage at half the cost. It just decimated these areas.

Danny attributed the "decimation" of Black neighborhoods to a history of disinvestment, but the frame of disinvestment is not the only one inspectors use.

Although Eddie was one of the most outspoken inspectors when it came to negligence and greed, he lamented individual as well as structural factors as we sat in his car in an African American neighborhood one morning. Like many others, this neighborhood has experienced redlining and institutional abandonment, followed by predatory lending. I watched Eddie's eyes casually following some children playing on a stoop, as he explained what he thought caused the prevalence of boarded-up buildings:

> A lack of education, the atomic bomb had been dropped on the family structure, which caused the grandma to become head of the household to begin with, the daughters and sons of said grandmother producing innumerable children, when grandmother died it was a place of congregation and . . . when they lived with their grandmother . . . she dies, there's no will, nobody has a plan. So, in many cases it becomes a party house, a drug house, again lack of education, lack of goals, lack of any ability to get or obtain a job, because there are no jobs in the local area.
>
> So, property taxes go into arrears, it takes years before the county takes possession of any of these buildings from a tax point of view. In fact, they just don't. They don't. What are they going to do? What tax investor is going to come in

here and buy these buildings for taxes? Because the appraised value of the property is below the amount for the back taxes. That's one of the calculations they use. If it's $10,000 on back taxes and it's going to cost me more to buy the place; or if there's $30,000 in back taxes and I can only sell it for $25,000. Forget about it!

While he was cognizant of structural explanations for dilapidation in Black neighborhoods, Eddie drew on racial stereotypes about African Americans. He mentioned the role of structural factors, such as unemployment, disinvestment, education, and property taxes. But he also pulled from a long history of ideas about Black family disorganization and common assumptions about a culture of poverty among Black communities. Because he recognized structural issues, Eddie likely did not think it problematic to also mention broken families, female-headed households, many children, drug use, and a lack of motivation. But individualized explanations for poverty obfuscate the structural nature of racism and racial inequality.

Flitting between structural and cultural frames was the norm among inspectors, even inspectors of color. I joined inspector Natasha for an inspection on a block of multi-unit greystone buildings dotted with boarded-up windows. The gate was padlocked, meaning we were unable to inspect the condition of the interior of the building we had been called to. Natasha, who carried a clipboard and adjusted her glasses frequently as she spoke, did not seem surprised that no one was around. We waited for a few minutes on the off chance that someone would turn up, and this gave us time to talk.

Natasha, one of only a few Black female inspectors, lives in a neighborhood adjacent to where she grew up. The daughter of a postal worker, she began training in carpentry through a local union as soon as she finished high school. She lives with her two children a few blocks from her parents, who still live in the house in which she grew up. She told me she was glad to have put even a few blocks between her current home and her childhood home, and laments the high rates of unemployment and crime "over there." She did not say as much, but her family could have been one of those that settled in the area in the 1960s as Whites—and institutional investment—left for the suburbs. "All you got to do is make the playing field even," Natasha told me, holding her clipboard up to shield the sun from her eyes.

> And that means, the same jobs that you're entitled to, I should be entitled to the same job. If we had the same income that would balance out. So we could get good food and take care of ourselves. We ain't asking for no handouts.

Inspector Natasha paused for a moment before continuing, and this pause seemed to signal a shift in her reasoning.

> But sometimes you go into a distressed neighborhood, and you just act out because that's all you see. It's no excuse but it's . . . the glass window or the broken window or something . . . if this is all you see, this is what you do. I went to [a middle-class White neighborhood] for work and I saw some kids out playing. They can't play like that in my neighborhood.

Natasha began by referencing disparate job opportunities and racial income gaps. But a secondary frame seemed to coexist in her mind as she discussed neighborhood inequality. Like Eddie, she drew on cultural explanations in addition to structural factors. Natasha mentioned the concept of broken windows to help account for racial differences in neighborhood conditions and safety. "Broken windows" is a criminological theory, and proponents of the theory argue that minor issues (i.e., broken windows) provoke more severe issues, such as crime and widespread neighborhood disorder. The underlying assumption is that minor issues left unaddressed signal a lack of collective efficacy and community organization, and this encourages people to see the neighborhood as a prime location for criminal activity.[17] Broken windows theory thus emphasizes the role of individual action in the formation of the kind of dilapidated and dangerous neighborhoods Natasha referenced.

Natasha's explanation also makes clear that it was not only White inspectors that possessed, and drew on, cultural and structural frames. In fact, I did not detect any difference between the overlapping frames of inspectors of color and the White men they work among.[18] This is a testament to the pervasive and stubborn nature of racialized frames. Inspectors across the board know the deck is stacked against homeowners of color, and they have overlapping explanations for racial disparities: from disinvestment to a culture of poverty. As Leslie McCall argues, while structural and individual explanations for inequality often seen like opposing principles, people tend to think in terms of both.[19]

White Blind Spots

Inspectors' frames also overlap with what I call "White blind spots." While inspectors acknowledged that housing conditions in communities of color stem from racialized policies and practices, they failed to recognize the ways that Whites have profited from discrimination and wealth extraction, as well as the policies that allow and subsidize wealth accumulation from homeownership and practices that inflate the value of property in White neighborhoods. "Look," inspector Steve told me during an interview, "on the South and the West Sides . . . you're always going to find the worst problems." Steve is a White inspector in his sixties with a loud voice and a heavy Chicago accent. He continued:

> Today, it's pretty much racial . . . the Black, the Hispanic communities are probably the most devastated . . . but you could go back fifty, maybe a hundred years and it was another ethnic group that was in that same situation.
>
> You know, I saw my mom on Monday and where she grew up at [on the South Side], they didn't have any hot water and that was just the way it was. They had a coal boiler in the basement, and you had to go down there and put coal in the boiler, so. . . . It's probably what you'll find is the South and West Sides, that's probably where the greatest concentration of heat complaints, porches, and just the utter lack of any kind of maintenance of buildings.

Steve compared the contemporary South and West Sides to earlier moments in Chicago's racial history—moments when, before White homeowners left for the suburbs, these now disinvested areas were majority White. He was telling me that he believed Whites of previous generations occupied structural positions similar to people of color today. There is something to Steve's statement. European immigrants in the early twentieth century, for example, had limited options for housing due to discrimination. But, as Doug Massey and Nancy Denton made clear, it is a mistake to liken the position of African Americans to that of other racial or ethnic groups in the US, because of the unprecedented and unique character of anti-Black discrimination.[20] Thus, by making this comparison, Steve was conveying his White blind spot.

Steve's comments reminded me of something else Eddie had said as we sat in his car on a South Side street surrounded by boarded-up buildings:

> That generation of African American is dying off. Just like that generation of Americans are dying off. . . . People my parents' age, who would be in their mid-eighties to mid-nineties now, you know, they're dying. And what is left is all kinds of shit.

Eddie's statement that a "generation of Americans is dying off" suggests that he perceived these demographic, economic, and cultural trends as extending beyond the problems facing Black communities. His pessimism encompassed all Americans, indeed even his own (White) family. But the connection Eddie—and Steve—drew between contemporary communities of color, ethnic Whites of previous generations, and contemporary Whites ignores how the deck has been stacked in their favor through policies encompassing housing, taxation, and land use. In short, inspectors seemed unaware of the extent to which the deck is stacked in favor of Whites.[21]

White blind spots are different from being colorblind. While colorblindness means a person fails to see or notice race,[22] people with White blind spots do see race, but not in its entirety.[23] White blind spots disclose only half the picture, and obscure the relational character of racial domination. White blind spots afford insight into disinvestment and disenfranchisement without revealing the wealth that was reaped through these practices. As a result, inspectors underestimate the extent to which White property owners have been set up to profit, in relation to property owners of color. They see communities of color as precarious, but they do not see Whites as profiteers. They do not direct their stabs at justice toward Whites in the same way that they do for other profiteers—landlords, for example. Yet this does not mean that inspectors do not penalize Whites. They do because, due to inequities in wealth and resources, the landlords and homeowners that inspectors try to penalize most heavily are most likely to be White.[24] However, inspectors tend to penalize obvious wealth rather than average (White) wealth. This is in part because wealth is often hidden. It is in the bank, in stocks, in inheritances, or behind the fences of gated communities, for example. As such, the accumulation of wealth that Whites uniquely have is not as conspicuous as poverty. While movies, music videos, and academic books are replete with images of urban poverty in communities of color, US society has been sold the idea that *the average*—i.e., not conspicuous—American home is occupied by Whites.[25] To be sure, extreme wealth is conspicuous, but average White wealth is normalized. As Dianne Harris demonstrates, Whiteness and what we think of as "ordinary homes" grew up together.[26] Just as we have not until relatively recently examined Whiteness as a construct, we have not noticed and have instead left unspoken the ordinary, nondescript homes that so often are a manifestation of White wealth.

"I Can't Walk Away from This": Race and the Obstinacy of Housing Conditions

Racial discrimination, wealth extraction, and institutional disinvestment mean that the deck is stacked most firmly against minority homeowners.

It thus follows that inspectors make stabs at justice by being lenient to homeowners in communities of color. But the numbers tell a different story. Quantitative analysis of the data set I compiled using building violation data and demographic survey data suggests that inspectors record more building violations in communities of color than in Whiter areas.[27] And this pattern is not just because there are more 311 service requests in communities of color. The geographic pattern in violations holds even if we account for the number of 311 requests, which means that the high number of building violations in Black, Latinx, and Asian neighborhoods is not just a result of high volumes of requests. Nor can this pattern be explained by inspectors' varying explanations for poverty, or their White blind spots.[28] So, what does explain this pattern? The answer lies in the obstinacy of racial disparities in housing conditions.[29]

A Black woman in a nightgown answered the door of a single-family frame house in a historically disinvested and majority African American neighborhood. The house, which was adjacent to an empty lot, looked to be in a sorry state. The flaked paint made it hard to determine the color of the siding, and broken windows were taped over with faded pale green fabric. Inspector Danny and I were responding to a 311 request about rats, but the rickety front porch caught Danny's eye and he seemed to forget to even look for rats. He wiggled the handrail and bounced up and down on the porch a few times to demonstrate its creaks and instability. "This is going to be a [court case]," Danny said to me, under his breath. "I can't walk away from this."

The severity of the porch's condition meant that Danny had no option but to bring the building to court. Although inspectors have a good deal of discretion, some issues compel legal action. This is due to both instructions from supervisors and what I saw to be a genuine sense of morality: dangerous buildings need to be fixed before someone gets hurt. But Danny's lack of choice in this case helps to illuminate why—regardless of where their sympathies lie—inspectors record more violations in areas with more residents of color. Inspector Derek expressed a similar sentiment:

> It's a tough issue. I mean let's face it, as people have become poor and it's harder to maintain housing, there's no question that the quality of the housing is deteriorated. Then you have to decide, and I'd have to decide then on the job, okay, now what do we do? You know that's a nightmare. The other thing is then what? When you find these conditions that are so appalling, of course

sometimes you have no choice but to immediately.... There was a case ... I'll never forget it, that it was so horrible.... We had to immediately get the people out because it was that bad.

As Danny and Derek explain, the existence of serious issues means that inspectors are duty-bound to take action, even if they would prefer to protect the owners of the properties they inspect by not recording violations or not beginning legal action. And due to a long history of limited housing options and obstacles to wealth as well as contemporary financial barriers to home repair, homes occupied by people of color are more likely to have severe building violations that inspectors cannot overlook, such as the unsafe porch that Danny noticed.[30]

Inspectors may have no option but to send property owners in communities of color to building court, but they can exercise leniency in the court cases that ensue. As we will see in chapter 5, inspectors in building court work with property owners to prioritize issues and allow them months—sometimes years—to make repairs. But the fact remains that they record more violations in communities of color, in spite of their attempts to make stabs at justice. Structural inequality can be obdurate and can constrain actions.[31] Despite inspectors' compassion—even though they "give a shit," as inspector Eddie put it—the obstinacy of material housing conditions limits the actions they can take and restricts their ability to shape the city.[32]

Conclusion

Precarious ties to homeownership, racial disparities in housing, the inability to keep up with maintenance, the threat of displacement, public-private partnerships, and the housing crisis—these are all features of the stacked deck. And these are the features that motivate inspectors when they assess homeowners, decide between leniency and punishment, and take stabs at justice. While stabs at justice often entail not citing property owners for minor violations, they can also involve attempts to steer city initiatives and even out investment and profit. In changing neighborhoods, however, inspectors assist newcomers as well as old-timers. Amid their concerns about unfair profit, inspectors exaggerate and expand categories of precarity and profit. This is significant because we often think of people and places as occupying fairly static unequal positions within the stacked deck. But inspectors show us that perceptions of the stacked deck can change, and can include and exclude compassion and disdain for and against different populations and places. Inspectors' underlying logic does not waver—they maintain their valorization of hard work and ownership of modest homes, coupled with a disdain for greed—but the people who slot into their categories can vary.

Other distinctions of property, people, or places are blurry because of the complexity of inequality. We can all be the subject and object of inequality within multiple hierarchies. The owner of a two-flat who rents out her second unit as a short-term rental might be a source of annoyance for her neighbors or for organizations lobbying to preserve affordable housing, but the rental income might be the only thing that allows her to avoid foreclosure and stay in her home. A developer may be unpopular with local residents for planning a new building among historic homes, but may ensure that some units remain affordable and thus increase housing options for low-income tenants. Categories and distinctions change in other contexts too. What kinds of crime do the police overlook in one context but not another, for example? What aspects of healthcare disparities drive medical coders to check a box that means a patient's insurance covers a course of treatment? Are there facets of the economy that motivate behind-closed-doors decisions regarding eligibility for unemployment payments? Could the political climate prompt someone at a polling station to let someone vote without a correct form of ID?

This chapter suggests that patterns in discretion may not be just a matter of bias toward some people and places. Instead, we should look at how frontline workers' categorizations of the stacked deck help them to make sense of relational positions within changing contexts. But some aspects of the stacked deck are stubborn and much less malleable. Irrespective of inspectors' categorizations and stabs at justice, for example, persistent racial disparities in housing stymie inspectors' stabs at justice in communities of color. This tells us something very important: that stabs at justice are not good enough. Justice requires a robust, foolproof system where disparities in resources and wealth are officially measured, acknowledged, and blunted. Without this, housing inequality will continue.

But to really understand the persistence of housing inequality, we need to see what happens after inspectors make their decisions. Chapter 5 takes us from on-the-ground inspections to the floor of building violation court, and to millions of online records of building violations. We follow up on the decisions that inspectors make and see that some aspects of the legal system and the housing market have the power to stunt inspectors' actions, obstructing their stabs at justice and reproducing the stacked deck.

CHAPTER FIVE

Justice Blockers

Chicago's building court was just getting going for the morning. I had barely taken my seat as the clerk called the courtroom to order. A few things stand out in building court as compared to other kinds of courtrooms. In the courtrooms we are used to seeing on TV, clerks call out the names of defendants. In this courtroom, however, clerks call the addresses of buildings—and building code violations are the crimes. Buildings—and their owners—wind up in building court when they have serious issues, such as dangerous balconies and porches, no smoke detectors, or structural damage. Owners (and their attorneys, if they have them) face the judge, an attorney for the city, and a building inspector.

Another thing that is different is the duration of cases. It is normal for building owners to periodically attend building court hearings—at dates set by the judge on the recommendation of the court inspector—over the course of months, or years, as they gradually make progress on their buildings. Court inspectors are like regular building inspectors, except that, instead of following up on 311 requests, they visit buildings listed in building court cases to check on repair progress. In hearings, court inspectors provide an update on the progress that owners have made fixing violations. Building owners remain in court proceedings until the violations are fixed or inspectors recommend that they get dismissed. Like their colleagues who respond to 311 requests, court inspectors use their discretion to make stabs at justice. In the courtroom, inspectors decide which violations must be fixed and which can go unaddressed.

The first building on the docket that morning was a small brick bungalow in a Latinx neighborhood on Chicago's West Side. Things were changing in the neighborhood—its White population was growing quickly, and developers were replacing bungalows like this one with expensive condo buildings and rentals. An inspection of the building, almost two years ago now, had turned up two issues: an illegally converted attic and excessive construction debris—bits of lumber and siding—strewn across the yard of the bungalow, which was nestled between modern three-story buildings. The elderly Latinx couple who owned the bungalow nodded to acknowledge the court inspector. They looked frail. The woman clung to her husband's arm, and the husband walked with a cane. This was their fourth appearance in court over nineteen months, and they had only made a small amount of progress in dealing with the issues that had landed them in building court. They had tidied the debris in their yard, but they had done nothing to correct the issues with the attic.

"I'm not going to make them get a permit," the court inspector explained to me. "I'm sure they have much better things to do with $3,000." I'm sure they did too. Decisions like this save property owners money and hassle. This elderly couple had already spent hours of their time in the courtroom over the past two years, but at least the inspector's decision— backed by the judge—would mean they would not have to pay for a permit, hire an architect, or do any more work to appease the building court judge. The man seemed to lose his balance momentarily when the inspector vocalized his decision. His grip tightened on the handle of his cane as he steadied himself. He and his wife looked surprised, I thought, as well as relieved. But this relief may be short-lived. What they did not know at this point was that their issues were not yet over. Their brush with building court had ended, but the unpermitted attic still shows up on their property's record as an unaddressed violation and—as this chapter shows—risks devaluing their property.

Chicago's building court inspectors take stabs at justice every day to help property owners like this elderly couple. Inspectors do not insist that low-income property owners fix all violations, or they allow them plenty of time to get the work done. And in the context of building court, inspectors can be pretty certain which property owners are low-income or lack resources, because the city attorneys they work with access financial records. But as we will see, the stacked deck endures, irrespective of these stabs at justice. Property values drop—especially on already low-priced property—and dilapidation persists or worsens. As it turns out, the stacked deck also interferes when inspectors take punitive stabs at justice. Inspectors punish landlords and property owners by citing violations and mandating that they make repairs. But the unregulated rental market swallows up these stabs at justice—by allowing (and sometimes even forcing) landlords to hike rents, upscale buildings, and displace tenants. Thus, irrespective of inspectors' stabs at justice, the stacked deck perseveres: low-income property owners are denied wealth accumulation and the city becomes more unaffordable to low-income tenants.

There's an important caveat. The things that happen after building inspections and court cases—property depreciation, dilapidation, rent hikes, and displacement—are not the direct effects of inspectors' stabs at justice. Rather, they are the outcomes of legal, bureaucratic, and financial systems in which inspectors' work is embedded. I call these "justice blockers." We have already heard in previous chapters about the racialized material obstinacy of housing conditions and the lack of recourse inspectors possess when it comes to tenants. These conditions make it very hard for stabs at justice to be effective. In the context of code enforcement, justice blockers also include bureaucratic and technological features of record keeping and the contemporary housing market, as well as characteristics of the legal system of code enforcement. People's uneven access to resources is also a persistent justice blocker. Tracing justice blockers helps us to understand how the stacked deck persists *in spite of* stabs at justice. Justice blockers hinder stabs at justice for property owners, landlords, and tenants.

How Justice Blockers Hinder Stabs at Justice for Landlords and Tenants

In this section, we see how the relatively unregulated rental housing market obstructs stabs at justice. Irrespective of inspectors' punitive stabs at justice aimed at bad landlords, tenants—especially low-income tenants of color—deal with the risks of unsafe or unpleasant living conditions as well as increasing rents.

Habitability and Hopelessness

As we saw in chapter 3, inspectors routinely make stabs at justice by citing landlords with as many violations as possible. But justice blockers render these stabs at justice ineffective at unstacking the deck against renters. One reason for this is because landlords often simply ignore inspection reports. And if landlords do not fix issues in their buildings, renters are stuck with unsafe or unpleasant living conditions. The building code covers a wide array of issues that are associated with poor health and danger. Lead, mold, damp, pests, and energy insecurity are linked to asthma and other respiratory conditions, for example, and poor lighting, structural issues, and a lack of smoke alarms, carbon monoxide detectors, and other fire safety precautions can lead to injuries and fatalities.[1] Black and Latinx renters are more likely than Whites to live in buildings with issues. This means that these populations disproportionately face the health risks associated with unresolved building violations.

If building inspectors do not insist on repairs, then who will? Renters—especially those with few options—do not have much power to make demands of their landlords, despite their legal right (in almost all states) to exercise what is called the warranty of habitability. This warranty permits tenants to withhold rent if their landlord fails to repair issues of which they are aware.[2] But the warranty regularly fails to protect tenants, in large part due to judges' favorable disposition toward landlords in eviction courts. Judges side with landlords if their tenants are not paying rent, irrespective of housing conditions.[3] Moreover, evictions can be cheaper than the cost of fixing issues.[4] All of this means that tenants may be punished for exercising their right to withhold rent in buildings with issues. The threat of eviction is stark, especially for tenants of color, who face bigger obstacles to finding a new place to live than do White renters. In addition to discrimination from landlords as they screen tenants, racism—in the criminal justice system and the housing market—means a tenant of color is more likely to have a criminal or prior eviction record, both of which landlords can legally use to deny rental applications.[5] Many tenants end up stuck in unsafe or unpleasant housing conditions, with little reason to be optimistic about making very legitimate requests of their landlords.

Raising the Rent

But what if landlords did fix the violations that inspectors record? This too can have negative consequences for precarious urban renters. I matched building violation data with rents across Chicago to estimate how fixing and not fixing violations relates to the rents that landlords charge. I gathered information on rent prices for a sample of units across the city, as well as building violation records for these buildings. Of course, the location of a building—the neighborhood it is in, and whether it's close to public transit or other amenities, for example—makes a big difference in the price of rent. So I also found out the average rent in every neighborhood.

Then I used statistical analysis to predict the price of rent for any property in the city, accounting for the average rent for its neighborhood location and the number of violations on record. The overall idea was to be able to pinpoint the relationship between building violations and rental prices (I give more details about this analysis in appendix A).

Remember Eddie's reaction in chapter 3 as we inspected a building with orange moldy ceilings, sparking outlets, a dangerous back porch, and an intermittent electricity supply? Eddie was furious about the landlord's negligence and—upon hearing that she charged $750 for rent—told the tenants he would pray the landlord went to "landlord hell." Eddie wrote up twenty-nine violations during that inspection, ranging from a dangerous porch to clogged drains and sparking outlets. I drove by the building a year later and saw some evidence that several repairs had been done. Where there had been boarded-up windows, there were now new windows and painted frames, and the porch system looked new and sturdy. But these repairs come at a price. The tenants who called 311 would no doubt be pleased with the repairs (if they are still there). But based on my data, if the owner fixed all twenty-nine violations, the rent could be expected to increase from $750 to $898 in a year.[6] That's a $148 increase in rent, which could be crippling to many residents, especially in this neighborhood where 76 percent of renters live below the poverty line.[7] But fixing violations does not correlate with increased rents only in this neighborhood. The relationship between fixing violations and increases in rents holds throughout the city.[8]

To be sure, for some landlords—especially small-time landlords who regularly face financial challenges maintaining properties[9]—rent increases may well be the only way to recoup the cost of repairs. For others, repairs may be a handy excuse to raise rents disproportionately. But without regulation, landlords are unchecked in their ability to raise rents and pass the burden of code enforcement onto their tenants, because there is no cap on the amount by which a landlord can raise rent.[10] In the public setting of building court, landlords discussed increasing rents freely, alleging that it was reasonable and was their only way to cover costs. While big jobs like fixing up a porch system on a mid-sized apartment building can cost a minimum of $20,000, small things add up too. In court one afternoon, one landlord of several large apartment buildings complained about needing to recoup $6,000 he spent on new window screens because of a relatively minor code violation. Another landlord was left with little choice but to hike rents, he told me, after he had to pay for heat treatment to rid his building of bedbugs—something he did not think was his problem because a tenant had caused the issue. "The City must think we have a printer," he said with frustration as he packed up paperwork to leave building court, "that we just print money. How do they think we can afford this?"

Across Chicago, resolving violations in rental buildings is associated with an increase in rent. This is a worrying trend because it signals even fewer affordable options for renters. A lack of affordable housing is a key aspect of the stacked deck. Thus, we are faced with a paradox: inspectors penalize landlords because of the profit they believe they make from the stacked deck, but this does nothing to challenge the stacked deck structurally. Again, renters who have the fewest options and who are in the most precarious positions in terms of stable housing are the most likely to bear the brunt of rent hikes associated with code violations. Black and Latinx renters not only face discrimination in the rental market; they also, on average, have fewer financial resources than do their White counterparts. Plus, there is evidence that many Black and Latinx Chicago renters live in the kind of large, professionally managed buildings that inspectors pursue for violations.[11] For these reasons, renters of color are hit hardest and most often by rent increases. It is worth noting that housing voucher holders, of which there are tens of thousands in Chicago, may not face this issue because increases in rents for their units are absorbed by the government. But since housing vouchers are so scarce, most renters—even poor ones—do not have this protection. Furthermore, although affordable housing is increasingly hard to come by for middle-class residents in contemporary cities, their limited choices can also further constrain the choices of their lower-income counterparts. In the context of tight housing markets, middle-class renters crowd out affordable housing, and there are fewer affordable units for low-income renters.[12]

But there's something else. Not fixing violations may not mean that landlords will keep rents affordable. In fact, my statistical analysis reveals that unresolved violations do not lower rent prices in Chicago. Even though correcting violations is correlated with higher rents, the opposite is not true. Surprisingly, rental units with code violations are not cheaper, on average, than units in better condition.[13] By and large, tenants in Chicago can expect to pay just as much for dilapidated or dangerous rental units with code violations. This echoes research that documents the shockingly small difference in rents charged in high-poverty and wealthy neighborhoods.[14] Although substandard housing was in part responsible for the development of building codes, these codes have not proven effective in curtailing exploitation in the rental market.

What becomes clear is that there are no options available to inspectors to help tenants that do not also risk negatively affecting them. Even when inspectors sympathize with tenants, there is no obvious route to assist them. This is because, although inspectors respond to tenants' requests, code enforcement is directed at property owners and is punitive in nature. Justice blockers are also to blame. In this case, justice blockers include the absence of regulations in the rental market and the persistence and accessibility of eviction records. As a result of these obstacles, tenants pay the price even when inspectors try to penalize their exploitative landlords.

How Justice Blockers Hinder Stabs at Justice for Property Owners

In this section, we see how property owners fare after inspections and court cases. When inspectors do not insist that owners fix all violations, property markers devalue property. Owners and communities additionally face a cycle of disrepair, in which they are not rewarded for their investments. The section also demonstrates the unequal access to justice in court cases and the paradox of legal leniency, by which property owners suffer despite compassionate rulings.

Property Markers

I use the term "property markers" to mean any negative credential[15] that can be associated with a property on its record. Property markers vary by state or municipal authority, and in terms of what data are publicly and/or easily available. They include records of unpaid property taxes, foreclosures or bank sales, liens, and other claims that cloud property titles. They also include registration as a vacant property, scofflaw listing, and building violations. Property markers, like criminal records, designate engagement with the legal system and work as a technique of evaluation and classification.[16] While criminal records can prevent people from renting or working,[17] property markers can make properties hard to sell. They can also prevent owners from making timely mortgage payments or paying their property taxes, which frequently results in foreclosure. Unlike criminal records, however, property markers are often a result of inspectors' leniency. In the following sections, we see the detrimental effects of one kind of property marker—building violation records—for properties, residents, and communities.

It didn't feel like spring, but it was. Inspector Vince and I arrived—in warm winter coats and hats—at a three-story brick building on Chicago's North Side. The street was quiet, the homes unassuming. In fact, very little stood out to me about this building, and I noticed I was having a hard time writing down a description as I usually did for buildings during inspections. The neighborhood is majority White, and the average median income is well above the city average, yet I would not necessarily describe the area as wealthy. This is an example of the kind of moderate White wealth that goes unnoticed because it is inconspicuous. The buildings on this street were neither new nor opulent, but neither were they run-down. Things were a little different around the back of the building, though, where we were met with a very flimsy-looking porch and an addition to the first floor that seemed little more than a wooden shack.

Vince was a building court inspector, which meant he knew more about the building than an average 311 inspector. He explained that the building's owner—who he had previously met in court—lived in this part of the building, which was built without permits. "He'll have no choice but to fix up the porch," Vince commented as he surveyed the building, before craning his neck and adding, "and take care of some tuck-pointing." Technically, Vince should also have insisted that the owner either remove the addition or apply for retroactive permits for its construction, which would entail paying the City fees and hiring an architect. Vince spent a long time staring in silence at the building, before confessing that he was having a hard time deciding whether he should insist that the owner address the issues with the illegal addition. "After the tuck-pointing and porch work, do I want to hurt him more?" he asked. I was unsure whether he was talking to me or to himself as he continued: "I'm out to help people, not hurt people." A few moments later, Vince made his decision. He told me that, in cases like this, which were not dangerous, he only insists that owners do the minimum necessary work. He would not pursue the permit violation. As we have seen in examples throughout this book, inspectors make stabs at justice for owner-occupiers by recording violations but not insisting that they get resolved. This is exactly what Vince was doing.

Yet, documentation of the illegal addition would remain on record for the building, and Vince knew it. He told me that the illegal conversion violation would "remain in the system." "He'll either have to go through the expensive process of getting permits later or risk the decrease in value to his property," he told me, adding that the owner would need to apply for the permit if he wanted to sell, refinance, or get a loan on the property in the future. "Banks will require it," he explained, because the building's "title is clouded." Vince knew enough about the legal and bureaucratic workings of property and the housing market to comprehend the implications of his action. The property owner was damned if he did and damned if he didn't. And Vince was too. The parameters of the code enforcement system meant that the only way Vince felt he could help this property owner would result in a property marker on the record of this building. This was not a one-off. Time and time again, inspectors dismissed cases from court—to show leniency and give property owners a break—before all violations were resolved. Property markers are particularly significant because of the availability of information.[18] As Matt—an inspector for fourteen years—explained:

> They made available online all building violations, so somebody can just check in and they can just go online and they can verify the address of the building and see the building violations on it. And what that's done is that a lot of building owners, when they go to refinance, the banks look and they see them now. They don't have to go through a lawyer in order to get a copy of it. It's right there online.
>
> So then those got to be corrected before they'll even get the loan. And some of them may be real minor stuff, but it shows up as an open violation. But it could just be that there was a broken window at one point. . . . Just another thing to go through on your loan!

Matt laid out the ways in which online property markers—even minor building violations—can affect property owners. Inspector Malcolm went one step further, telling me he suspected that some people lodged 311 requests for the very purpose of creating property markers. Developers, he believed, would make 311 calls on buildings they were interested in purchasing to redevelop in lucrative areas. They would then be able to use the recorded violations "for leverage to get the price down." "It's hard to sell a building with violations," Malcolm went on, "banks don't want to give loans and buyers are very wary of buying." He believed property markers to be significant enough that interested parties want to cook the data.

Paying the Price

Are Matt and Malcolm correct? Do records of building violations take a toll on property prices? Are property markers as significant as inspectors, developers, and banks believe? I matched building violation data to data on property transactions and neighborhood sales trends in Chicago to try to find answers to these questions. These data gave me information about some key things. I had information about code violations that were on a building's record, I knew the sales prices of these buildings, and I also had data on the average changes in home price for every neighborhood in the city. Using regression analysis, I could then calculate a predicted price for any property over a five-year period. The predictions took into account the number of building violations a property had on record and what we would expect the price to be based on its location in the city and the year in which it sold (I give more details about this analysis in appendix A).

I discovered that unresolved building violations mean lower property prices. You might remember when, in chapter 4, inspector Nick opted not to insist that a homeowner resolve violations on her property because it was for sale and he did not want to "add to her headache." Nick recorded the twenty-six violations he found—ranging from broken windows to a dangerous porch and collapsing siding—but did not insist that the owner fix them before she sold her property. Although Nick was trying to help the owner, his strategy may have backfired. These violations—which lingered online, both in the city's building violation database and on the property's real estate listing—correspond to a $58,476 drop in price.[19] If the owner sells her building, she can expect to get $625,014 instead of the $683,490 she could have received if her building had no lingering violations.[20] Of course, most homes do not have twenty-six recorded violations. But even four unaddressed violations could be expected to correlate to a $9,340 reduction in property price. The owners of buildings marred by property markers pay the price.

Moreover, property owners with the lowest valued homes are disproportionately burdened by property markers. Although unresolved violations are associated with decreased values for all homes, on average they seem to have a greater effect on less expensive properties.[21] This trend has implications that reproduce the racialized hierarchy of the stacked deck. Legacies of discrimination mean that, on average, Black and Latinx residents have much lower property values than Whites,[22] are more likely to live in housing with building violations, and have fewer financial resources to make repairs.[23] Owning property is seen as a crucial motor for economic stability, mobility, and wealth. Yet even when people overcome the barriers to ownership, the benefits of homeownership can still be far out of reach because of property markers, which act as justice blockers and blunt the stabs at justice that people like building inspectors make.

Pros and Cons of Repairs

"Have you heard the broken windows theory?" inspector Danny asked me as we drove slowly through a dilapidated block one afternoon, veering from one side of the street to the other to avoid the occasional pothole. "You have to catch stuff before it escalates," he told me, "You have to get it at that point before it deteriorates." Danny was referencing the criminological theory of "broken windows," which holds that small-scale issues lead to big problems if they remain unaddressed. While this theory is hotly debated among criminologists, it makes a lot more sense when it comes to buildings: conditions get worse if left unaddressed. It doesn't take a building inspector to know this. Not maintaining buildings, or getting behind on necessary repair work, can lead to serious problems and dangerous conditions down the line, especially if issues are structural. Inspector Eddie told me something similar as he parked in front of a red brick building that he had previously written up for crumbling brickwork:

> Maintenance has to be done on a regular, annual basis . . . the three little pigs never talked about the fact that they had to tuck-point that brick building. With a wood building, you've got rotted wood sheeting underneath the siding. It's very expensive on both ends. . . . Brick is not more sustainable than wood. Wood rots . . . both are gonna start really being dangerous and hazardous.

Eddie pointed to the eaves of the building we were parked by, shaking his head. He was telling me that all buildings deteriorate over time. This is as true of single-family homes as of big apartment buildings, and it applies for both wood frame and masonry structures.

Even some issues that seem relatively minor—such as mold or damp— can be health risks. What's more, a minor issue overlooked might become a major issue in the years that follow. For example, even small holes and cracks in exterior walls can lead to severe structural damage if they are not repaired, because water easily gets into holes and forces bricks and mortar to crumble and pull apart. In fact, over fifty violations that inspectors regularly record are issues that are likely to get worse if left unaddressed. Thus, stabs at justice that entail overlooking minor issues can become big problems for property owners in the long run. This catch-22 situation could be avoided if property owners had funds to make repairs.

However, making repairs may not pay off for property owners. In fact, fixing violations generally has no effect on a property's sale price.[24] According to my analysis of building violation data and property transactions in Chicago, repairing violations—from broken windows to collapsing roofs—is not linked to increases in property prices. This means that if the owner of the building with twenty-six violations fixed them, she should not expect to see any financial return for her efforts. While fixing violations might improve dangerous conditions, it does not, on average, correlate with higher property value. This seems surprising, but it makes sense if we go behind the scenes of the housing market and understand that the real estate industry determines the value of a property by the value of the land as well the value of the building. And land values are derived from attributes of a location, such as reputation and proximity to amenities.[25] As such, land values can decrease and cancel out any effect that renovations and repairs might have on a property. Even if property owners pay out large sums of money to make repairs, they cannot expect a return when they sell their homes. While repairs may address the health and safety risks presented by building violations, property owners are not financially rewarded for making repairs—only penalized for disrepair.

The housing market has long been a source of racial and economic inequality in cities. That influence continues. Inspectors are limited by the tools at their disposal for making stabs at justice, and property owners have no good recourse in the context of the broader housing market, irrespective of inspectors' decisions. Stabs at justice are also stymied in building court.

Repeat Players and One-Shotters

The simplest way for a building court case to be dismissed is to complete the repair work necessary to clear a property of building violations. Yet this seldom happens; owners rarely complete all the required repairs. Instead, as we have seen, inspectors recommend that cases be dismissed when they are satisfied that enough work has been done. To an average property owner, it may not be obvious what exactly needs to be done. Some property owners, however, know the ins and outs of building court. "No, no, we want to comply with all the violations," came the surprising remark from a property owner in building court one morning. The court inspector and city attorney looked baffled. They were used to property owners fighting tooth and nail *not* to have to fix the building violations that brought them into building court. And, as we have seen, building violations cases are commonly dismissed with substantial compliance, meaning minor violations are overlooked once severe issues are addressed. Yet this property owner—who was planning on selling the two-flat he had rehabbed without obtaining the necessary permits—knew that any violations he did not address would remain on his building's record. "I'm going to flip it!" he said with enthusiasm. "And you can't usually get a mortgage with violations [on record]," he explained, thinking ahead about the implications of violations for prospective buyers. "People don't want to see anything on their title." The knowledge and confidence of this property owner stood out. As a self-confessed property flipper, he was, in the words of Marc Galanter, a "repeat player."[26]

The legal system rewards "repeat players"—those who know how to play the game, have financial and legal resources, and whose livelihoods are not threatened by litigation.[27] Repeat players win at the expense of "one-shotters," who have little in the way of legal expertise or resources and are in precarious positions such that litigation threatens their livelihood. In the context of building court, repeat players are often professional landlords or property flippers who know the system or have legal representation in building court.[28] As one landlord told me, scanning the coffee shop to see who might be listening: "Bad building owners take advantage of the time they get in court. They know how to play the system." Knowing how to play the system—for example, by avoiding property markers—means that repeat player property owners can ensure that building court does not stack the deck against them.

One way that courts and legal systems typically affect defendants is through fines and fees. Chicago's building court only occasionally imposes fines—usually only if cases have dragged on for many years. For example, a judge dismissed one property owner's case with a $500 fine "to cover all of this we've been through over these years." "That's a bargain," the inspector added. Bargain or not, fines are a source of further hardship for low-income property owners. As with other monetary sanctions, building court fines are regressive. Even if fines and fees are imposed on everyone equally, they have the most deleterious effect on those with the fewest resources.[29] Disproportionate effects are unfortunately no surprise in legal settings, but in building court they even extend to situations where courtroom actors attempt leniency.

Hassle and Progress

Court inspectors often give property owners more time to address concerns, seeing such extensions as stabs at justice. But more time exacerbates what Issa Kohler-Hausmann calls "procedural hassle."[30] Procedural hassle entails delaying, engaging, and compelling the defendant to conform to the institutional and organizational demands of the court and court actors. Quite often this simply means showing up for court, which entails lost wages because of missed work, additional childcare costs, attorney's fees, and wasted time.[31] I saw this happen over and over again in building court.

A White woman in her twenties hushed her baby in its stroller as she related her story to the court inspector. Like many others in building court, the woman's family property had a porch that desperately needed repairs. Some of the steps had rotted and the handrails were loose. Her father, who had taken care of the building, had died a few years ago and her mother was housebound. This was her second court date, and she had promised to have made progress on the porch by now. She showed the inspector photos on her phone of some work she had completed, including removing some debris and fixing some door glass. But nothing had been done to the porch. The inspector looked at her. "Your family has owned the building for a long time, right?" he asked, and watched as the woman nodded. He agreed to allow her more time and set her another court date four months later. "That's a long time," he said to her, "so you should be done by then." If he had not granted the woman more time, she would have had a choice to make: find the funds to repair the porch immediately, or allow a third party to do the repairs on her behalf. The latter option could mean this third party (i.e., a receiver) would be able to take control of the property. With these alternatives in mind, it's clear that the inspector was showing leniency—attempting a stab at justice—by allowing the woman additional time to fix the porch.

This woman was not the only beneficiary of inspectors' courtroom leniency. Tom—a sharply dressed Asian man in his thirties—was another. This was his building's third court date, and a different defendant had appeared at each hearing. His mother and aunt had been before—whoever could get the day off, Tom explained. The two-flat building needed a new roof and a new coat of paint. Tom told the inspector he had started on the paint job while he was trying to amass funds to fix the roof. The inspector made it clear that he appreciated their cooperation, and gave him "a long date," meaning Tom, or another one of his family members, would not have to come back to court for six months. Tom seemed pleased as he left. But the extra time inspectors afford to struggling homeowners as a form of leniency is not always beneficial. Extra time means extra hassle. Procedural hassle is particularly significant in building court, because some cases last for years (the longest I know of is eight), with intermittent court dates usually every three or six months at which property owners are expected to report on their progress.

"Progress," a building court judge told me, "that's what judges want to see. This is the thing that's very important. Progress. They want to see progress . . . everybody knows that our ultimate goal is to get the repairs done, as long as there's continual progress." Showing progress grants property owners more time in court, which, to building court judges, attorneys, and inspectors, is a blessing, especially for low-income owners. "[We're] aware that sometimes, along the way, people run out of money." The same judge told me,

> They need a little bit of time. If you can demonstrate that all the way along the way, you've been working your way towards the goal, then [we're] going to be more prone to saying, "You know what, you ran out of money. You've done this much. Let me give you a little bit of time so you can raise a little more money, a little more capital, and continue going." As long as [they] can show some kind of progress, and movement in the right direction, that's what [they] want to do.

To get their case dismissed from building court, defendants must show "progress," make repairs, or acquiesce to a court-ordered third party making the repairs. In building court, repair work is also subject to more scrutiny than it would be otherwise. Property owners must provide evidence of permits and plans, and they must be able to show that their contractors, plumbers, electricians, etcetera are licensed. Often the court requires property owners to hire an architect, which can prove costly. While these steps and pieces of evidence are technically required for repair work that is not mandated by the court, property owners are largely unchecked outside the courtroom. Property owners make repairs; renovate buildings; build garages, porches, and additions; and erect fences without obtaining appropriate permits or hiring licensed professionals all the time. A court mandate—which is the result of leniency in this context—brings defendants under the watchful eyes of the court.

Not only must defendants spend precious minutes and hours waiting in long lines or sitting in the courtroom looking confused before their hurried attorneys turn up late from another hearing across town and judges return from long lunch breaks. More significant, perhaps, is that defendants in building court wait months between hearings, and often years before dismissal. Sociologists argue that waiting is a powerful way that marginalized populations are conditioned to understand their subordinate place vis-à-vis government bureaucracies and services.[32] While most studies document waiting in person (in welfare or immigration office waiting rooms, for example), waiting at a distance between building court hearings may also communicate to property owners—who are disproportionately low-income or unable to afford repairs—that they are objects of scrutiny and are at the mercy of the courtroom bureaucracy.

The Paradox of Legal Leniency

The status of many homeowners of color, coupled with inspectors' leniency, creates a paradox. Inspectors are more likely to go easy on homeowners of color in court because they are more likely to be struggling with the burden of home maintenance. But this means that these homeowners are also more likely to end up with property markers or dilapidated buildings, and thus to lose value on their homes. Justice blockers transform stabs at justice—attempts to help property owners inspectors deem to be struggling or low-income—into tools that exacerbate the stacked deck. Overall, whether inspectors go easy on property owners in communities of color by overlooking violations or by recording violations and showing leniency in building court, property owners are saddled with either dilapidated property or property markers or both.

So what happens when, instead of leniency, court inspectors opt for punitive stabs at justice? The short answer is that property owners face negative ramifications from these actions as well. This was evident on the occasions when I saw inspectors' patience run out. The owner of a two-unit frame house in a majority White North Side neighborhood had failed to show up in court, and the inspector and attorney were filling out some paperwork on a coffee-stained table in front of them. The property was in court because the porch was—according to the inspector's notes—in "poor condition." "Would you consider the porch dangerous?" the attorney asked. "No," the inspector replied as he glanced at his own paperwork, seemingly to refresh his memory. The attorney's pen had not touched the paper before the inspector changed his response. "Put 'yes,'" he said simply. "Put the porch down as dangerous. [The owner] is not making progress, so let's say dangerous to make it a shorter date." Giving the owner a shorter date meant less time to repair or make progress on the porch. Punitive actions also result in unfavorable outcomes for homeowners who do not have money at their disposal.

Inspectors have the power of postponement to help building owners, but they can retract that help at any time. "It leans so much it almost falls over!" court inspector Tyler told the judge about a three-story building on the North Side. But that wasn't the only problem. The property had been in court for years because of severe structural issues, and the owners had failed to turn up for their two most recent court appearances. They were here today, but had no progress to show. "You have run out of time," Tyler curtly told the owner, a casually dressed Latinx man with gray hair. "You can plead your case with demo court," he added as the owner opened his mouth to reply. Tyler was recommending the building for demolition. I knew this was a big deal. Des, an inspector I had interviewed, had explained the demolition process to me a few days earlier. Property owners are liable for the costs of demolition, which can run up to $20,000. Very often, Des told me, owners cannot afford the expense, "so what'll happen is, we pay the contractor up front [to demolish the building] and lien the property." The building court inspector's decision—his lack of compassion—would end up costing the property owners dearly.

Disparate Justice

Building inspectors and other frontline workers take stabs at justice, but the broader legal and bureaucratic fields in which they work are justice blockers—they are already set up to benefit certain players. Every law has tension built in. The building code is intended to provide safety, but it also metes out punishment through the financial burden of citations and property markers. The reality of building court means that both punishment *and* leniency have negative implications for property owners without resources because of how the system is set up. The legal and bureaucratic framework of building code enforcement often bolsters the status quo because face-neutral laws that do not discriminate on the surface have disparate effects.[33] In fact, laws and regulations are often limited in their ability to counteract inequality precisely because they are neutral, whereas the people on whom enforcement acts occupy different positions and have disparate access to resources. Those with status and wealth can evade the costs and burdens of enforcement, and those without cannot. Justice blockers also mean that this happens in spite of efforts to assist people in marginalized positions. Justice blockers recast stabs at justice as injustice and exacerbate the stacked deck.

Justice blockers can also prevent people from attempting stabs at justice by making stacked decks seem insurmountable. Many people faced with stacked decks resign themselves to the status quo. In such cases resignation is not a product of delusion or an unawareness of inequality, but a result of change being risky and hard to achieve. In other words, the problem is not that people are unaware of stacked decks, or complacent about them. Rather, it is the stacked decks themselves, the justice blockers they comprise, and the way they leave people feeling powerless to make a difference.[34] What's more, justice blockers blind external observers to stabs at justice. We can't see them unless we really pay attention. Take the case of building code enforcement: we see seemingly constant dilapidation and dangerous buildings in poor neighborhoods, and we assume this means building code inspectors do not care. The reproduction of urban inequality obscures the motivations and intentions of inspectors. Code enforcement appears to reproduce the status quo, even though inspectors often break with the status quo. When we lack understanding of inspectors' motivations and actions, their actions appear to directly reproduce housing inequality. In fact, the stacked deck is reproduced in spite of, not because of, inspectors' actions.

The housing market and the legal system are not the only structures that obstruct stabs at justice. Welfare reform, for example, limits case workers' options when they attempt to guide their clients through the increasingly complex world of state assistance.[35] Policing tactics end up thwarting well-intentioned actions on the part of police officers.[36] Frontline work is likely replete with justice blockers that we have yet to uncover. Do regulations about occupancy limits or restrictions on substance users at homeless shelters prevent service providers from admitting the vulnerable people they want to house? Do policies about standardized tests or policing in schools obstruct teachers from making effective stabs at justice? And to what extent do justice blockers make us feel powerless? We know there are persistent problems, from under-appraisals of property owned by people of color to huge gender pay gaps—but the intransigence of these issues overwhelms us to the point that there seems to be no way out. How do justice blockers blind us to stabs at justice taken in these and other settings? We may miss the stabs at justice of mortgage brokers, for example, due to the overwhelming and complex characteristics of financialization and assessment that obstruct their actions, stymieing their stabs at justice and rendering them invisible. Thinking about these actions and contexts in terms of stabs at justice and justice blockers helps us to see what people must navigate in the settings in which they operate, and helps us to rethink the relationship between actions and effects.

In fact, this chapter calls for a broad reconsideration of the use of the word "effects" to describe what happens after decisions and actions. We need to move away from this term because it implies that events unfold either intentionally and straightforwardly from actions, or as unintended consequences. In fact, events are set in motion by justice blockers that envelop discretion and recast stabs at justice as injustice. Consequences are often neither wholly intended nor unintended; in fact, this distinction is not particularly useful. What is more important is that consequences may be unintended but not unexpected, given what we know about the social and political structures in which actions are embedded.[37] Frontline actors, and people more broadly, have discretion, but the motor is elsewhere. The concept of justice blockers helps us to see this and to explain the persistence of the stacked deck. Where does this leave us? To effectively mete out justice, we need to acknowledge disparities in resources[38] and seek to close the gaps.

Conclusion

Taking someone down a peg or two. Nitpicking. Tying someone up in red tape. Doing someone a favor. Turning a blind eye. These are the tools at the disposal of frontline workers. They are acts of leniency and punishment, and—more often that we think—they are stabs at justice. But despite frontline workers' stabs at justice, the stacked deck persists. It does not persist as a direct result of frontline workers' decisions and actions, though. Nor is its tenacity an unintended consequence of stabs at justice. Instead, as this chapter has shown, justice blockers prevent stabs at justice from effectively challenging the stacked deck. Characteristics of the contemporary US housing market—from stipulations on loans and refinancing to policies that mandate the availability of digital information about building code violations—widen gaps between poor and wealthy property owners. The lack of regulations in the rental market and the punitive nature of code enforcement are a toxic mix that renders inspectors' discretion part of the problem of a lack of good-quality affordable housing. Features of the legal system—such as the disparate impact of fines, disparities in legal resources, and procedural hassle—further exacerbate inequality among property owners. Although building inspectors in Chicago try to shape the city according to their perceptions of injustice, the stacked deck prevails due to the realities of the housing market, the legal system, and entrenched racialized poverty.

Refracted through the prisms of legal, financial, and bureaucratic features of the housing market, stabs at justice simply cannot disrupt the stacked deck. Put more generally, justice blockers ensure that stabs at justice fail, and that the stacked deck persists. But this is contingent on the presence of justice blockers, and is not inevitable. The book's conclusion takes up this critical point and offers some suggestions for how we can avoid the impasse and reshuffle the deck.

CONCLUSION

Reshuffling the Deck

The cities I know best—Chicago, New Orleans, and London—look quite different. Chicago is known for its neighborhoods of brick buildings and the skyscrapers that dot the city's famous downtown skyline. Most homes in New Orleans are made from wood. The city's characteristic shotgun-style homes are long single-story structures that are often raised to maximize ventilation and protect homes from flood water. In London, buildings abut one another on narrow streets—from grand-looking Edwardian buildings to twentieth-century tower blocks. The residential landscapes in each city also tell stories of disparity. Chicago is known to be a city of neighborhoods, but it is also a city of neighborhood differences. Rows of empty lots and vacant buildings characterize some neighborhoods, while others comprise luxury residential developments built among historic greystone homes. In post-Katrina New Orleans, brand-new two-story buildings make their neighboring structures—small, dilapidated, and overgrown with jasmine—look like shacks. Meanwhile, absentee owners buy up, but rarely inhabit, London's housing stock. Even empty buildings displace people. Working- and middle-class Londoners are left with cramped and increasingly squalid flats whose walls hold decades of mold and damp.

Day-to-day inequality hits home when disasters happen. It is no coincidence that Black New Orleanians were disproportionately unable to return and rebuild after Hurricane Katrina. Likewise, it is not happenstance that the seventy-one people who died in the Grenfell Tower fire in London in 2017 and the ten children who died in 2018 in a small rental building in Chicago were members of populations with the most precarious ties to housing: low-income renters, immigrant families, and residents of color. It is also no coincidence that house fires are concentrated in the poorest neighborhoods in US cities.[1] Across the world, buildings burn down or collapse; homes are ruined, and lives are lost. And homes in communities of color are consistently most at risk.

And what do we do about it? The media rehashes well-trodden narratives that shame municipal governments for failing to enforce building codes. Governments usually proceed to point the finger at other urban actors: landlords, architects, developers, or contractors. Independent investigations and litigation regularly ensue, and sometimes lead to an adaptation to building code regulations. Such has been the piecemeal pattern of building code enforcement over the last century and a half. This is not getting us very far: building code regulations are not successful in preventing disasters that disproportionately affect marginalized residents. Nor do they improve housing conditions for low-income tenants or communities of color. The deck has been stacked against these people and places, and it can seem impossible to move beyond this impasse.

But it's not impossible if we listen to the people who have to deal with this impasse every day. We have a host of evidence to suggest that the system of code enforcement does not work, but we cannot really know how or why without hearing, watching, listening to, and learning from those who navigate the system. A multitude of people—from building inspectors, bus drivers, insurance agents, landlords, and medical coders to police officers, prosecutors, and judges—work within and against the uneven places and systems that sociologists theorize. We need to know what these frontline workers are up against. What obstructs their attempts to help people? What makes it easy for them to punish people? How do they make sense of the contexts in which they work? Learning lessons from the front line is the best way to reveal a path through the impasse. We need to see how the things people have to deal with get in the way of achieving change. This is the most effective way to find pragmatic solutions to pervasive issues. In taking up this task, this book posits three concepts: stacked decks, stabs at justice, and justice blockers. Taken together, this conceptual apparatus points us toward workable solutions that can reshuffle the deck.

Stacked Decks

Stacked decks—hierarchies of institutionalized inequality—are everywhere: from our schools, neighborhoods, and places of employment to systems of criminal justice and healthcare. Overlapping policies, trends, and actions build up each stacked deck over time. Although they can seem insurmountable, stacked decks motivate people to act. They can be a rallying cry. In this book we have seen how the stacked deck motivates one group of people: building inspectors in Chicago. As frontline agents with discretion, they need to slot people, property, and places into categories. Though their job is to assess the condition of buildings, they use manifestations of the stacked deck to make their decisions: poverty and precarity, luxury and profit, dilapidation and disinvestment in communities of color, negligent landlords and struggling homeowners, the legacy of the 2008 housing crisis, public-private partnerships and other government initiatives, and upscaling and neighborhood change. Inspectors use the stacked deck to forge categories, and these categories guide their decisions and help them do their jobs. While most existing accounts assume that urban regulatory actors act in concert to further economically uneven urban development,[2] this book shows that some city workers oppose relentless attempts to make land more profitable at the expense of low-income and minority communities. Inspectors diverge from the priorities of municipal governments and have their own categorizations of the city, based on their perceptions of how the deck is stacked.

Stabs at Justice

Dragging people through red tape, strictly following and enforcing the rules, overlooking a violation, bending the rules, helping someone maneuver around the rules, finding alternatives, hurrying a case. These are the tools of frontline workers when they make stabs at justice—small and immediate acts that are aimed at leveling the playing field in response to perceived injustice. Stabs at justice comprise the tools at their disposal—from discretionary decisions and subjective interpretations of regulations, to selective enforcement, doctoring paperwork, and imposing waiting periods and delays. In this book, we have seen multiple examples of inspectors' stabs at justice. Motivated by the stacked deck, at times they overlook issues, favorably bend the rules, give people more time to fix issues, do not require additional steps, and help property owners benefit from the rules. In other situations, they are sticklers for the rules: they nitpick, are overly thorough, and record as many issues as possible; they are unhelpful, insist on following the letter of the law, and give people little time to make repairs. Each of these actions is a stab at justice—situational, emotional action, motivated by anger, unfairness, and injustice.

Stabs at justice are also about picking battles and opting for pragmatic actions aimed at immediate goals. Therefore, they can appear short-sighted. The concept of stabs at justice builds on Scott's theory of power and resistance and the actions that exist between explicit resistance and total compliance.[3] When we think about frontline decisions as stabs at justice, we can learn more about how inequality is reproduced, and better understand the scope and limits of frontline workers' actions.

Justice Blockers

"Everyone hates building inspectors," inspectors often told me. They may be right. Landlords think inspectors are out to make money for the City, tenants think they are in cahoots with developers and management companies, and homeowners think they are nitpicking and punitive. Even the landlord of my well-maintained apartment building in Chicago fears inspectors. "Don't bring them here!" she cautioned when I mentioned my research. The concept of justice blockers helps us to understand where these sentiments come from and suggests that landlords, tenants, and property owners disdain building inspectors because they are easier targets than, or are assumed to be representatives of, the justice blockers that actually dictate the fate of buildings in the city.

Justice blockers prevent attempts at justice from being realized, obscure the existence of these attempts, and make compliance with the status quo seem the norm. In the context of frontline workers, justice blockers prevent the selective enforcement of laws and regulations from having redistributive consequences. While building inspectors make stabs at justice, aspects of the housing market and bureaucratic and technological characteristics of the legal system corral inspectors' actions and end up reproducing the disparities that make up the stacked deck.

Justice blockers also explain why little seems to change, irrespective of good intentions. White families have nearly ten times more wealth than Black families and are much more likely to live in good-quality housing in well-resourced neighborhoods. Much of this disparity has been attributed to housing policies that have historically favored Whites.[4] But disparities persist even when those who implement policies try to protect communities of color. In fact, intentions—good or bad—are less important than the structures in which well- or ill-intentioned people operate. Thus, the politically expedient task becomes tracing how stabs at justice are turned into injustice. We need to identify justice blockers if we are to reshuffle the deck. To do this, we need to stand in the shoes of frontline workers.

Standing in the Shoes of Frontline Workers

Mindless bureaucrats, paper pushers, cogs in the machine. The terms we use to depict frontline workers downplay their judgments and blind us to how aware they are of the stacked decks in which they work. Yet frontline workers' capacity to make change and reshuffle the deck is limited. Discretion in frontline work is unavoidable, but we cannot rely on it for justice. Even if inspectors' stabs at justice successfully created an informal form of redistribution among homeowners, for example, this would not amount to codified and legally institutionalized policy. Homeowners would be unable to call on or expect official assistance.[5] In fact, "successful" discretionary stabs at justice like this might undermine the need for codification or government intervention. Relying on discretion is not good enough.

Doing away with discretion is also not a good idea, as Virginia Eubanks makes clear in her book *Automating Inequality*. In fact, automating frontline work does more harm than good. Eubanks describes how automated decisions—arrived at using algorithms, data mining, and computer models that predict risk—about whether unhoused people are eligible for services and who qualifies for unemployment assistance or healthcare create "a feedback loop of injustice" that further stigmatizes and penalizes poor and working-class citizens.[6] Furthermore, by removing face-to-face interactions, automation makes decisions and outcomes harder to contest. Yet automation is all the rage. City governments are joining forces with data and technology companies to tout the potential of "smart buildings," which predict building code vulnerabilities, among other things. There is huge financial backing for these projects nationwide, and their professed goals range from increasing property values, cracking down on code violations and discerning which buildings are more vulnerable to fire, to improving energy efficiency.[7] "Smart buildings," however, use existing city data, collected and assembled by frontline workers like building inspectors. As this book has shown, these data are far from objective assessments of building conditions. Instead, they are evaluations of people, profit, and precarity.

Using them as mere indicators of the physical condition of buildings is unlikely to yield accurate predictions.

Increasing the number of inspections is not the answer either. In Chicago, as in other cities, tenants' organizations campaign for proactive rental inspections, in which all rental properties must pass an inspection before they can be occupied. But without changes to justice blockers, proactive rental inspections are only likely to exacerbate current issues. If inspectors continue to punish professional landlords, proactive inspections will simply mean an increase in rent hikes and additional displacement. Similarly, we should be cautious if building codes are, as climate advocates and architects propose, targeted as a means to reduce carbon emissions. If inspectors continue to enforce codes more stringently for wealthy property owners, this may mean that already resourced residents get to live in climate-friendly buildings while low-income property owners are left living in buildings that are increasingly labeled as problematic. Finally, while corruption is a serious issue, paying it too much attention risks obscuring the broader issues that make building codes hard to enforce: a lack of resources and unregulated profit.

Changes to code enforcement, the building code, or the volume of inspections are not, by themselves, effective ways to address the stacked deck. Where does this leave us? The change needs to be at the level of justice blockers and resources. The following paragraphs review some policies and interventions currently aimed at reducing housing inequality—for both renters and homeowners—from different angles. While some are more readily doable than others, the persistence of inequality demands that we equip ourselves with as many tools as possible to tackle justice blockers.

Reshuffling the Deck for Renters

We need housing regulations and building codes to ensure safe and decent building standards. But, without a tandem system of rent control in place, landlords are unchecked in their ability to raise rents and pass the burden of ensuring safe housing—specifically, the costs incurred from compliance with building codes—on to their tenants.[8] Rent regulations could only allow rent increases in buildings without code violations. But we also need a system to protect tenants from rent hikes to cover costs of repairs. Governments could subsidize repairs on the condition that landlords provide affordable housing. Small rental buildings remain a significant source of affordable housing, so rent control regulations would need to avoid impoverishing small-time landlords. Small-time landlords could be included in an expanded system of municipal or federal grants or low-interest loans to property owners for repairs.[9] The question of affordability is also contingent on renters' incomes. Ensuring affordable housing thus also necessitates joining forces with unions and labor organizations to fight for living wages.

Housing choice vouchers offer another alternative. The housing choice voucher program, which currently covers tens of thousands of Chicagoans and over 5.3 million people nationally, absorbs rent hikes. Housing voucher recipients pay 30 percent of their income toward rent. Thus, if a landlord makes repairs and increases rents to cover costs, tenants are shielded from the rent hike.[10] The voucher program may do little to deconcentrate poverty,[11] but we could build on the level of protection it offers and expand it to tenants of all incomes. Cities could also take and exercise additional power to intervene when property owners abandon property, perhaps—as Hackworth suggests—fixing them up and gifting them to landlords with good track records who commit to keeping units affordable.[12] These suggestions underline the importance of government intervention to ensure affordability. Interventions should also detach the provision of affordable housing from profit. Not only does profit-seeking often lead to exploitation, and widens gaps between the rich and the poor, it is an unsustainable model, especially in crises and volatile markets.[13]

Affordable housing also needs to be safe. But too often, there are months-long lags between inspections that uncover dangerous conditions and court case rulings that deem rental buildings unfit for habitation. These lags can be deadly. Vacating buildings sooner and allocating more resources to building court to facilitate immediate hearings would also mean slumlords could not collect rents while waiting months for their court case and leaving their tenants in limbo. While on-the-spot vacations would only work if the city had and maintained a stock of decent options for temporary housing for displaced tenants,[14] this proposal would also give inspectors a recourse for dealing with tenants which they currently lack.

Similarly, we can do more to ensure that warranties of habitability are implemented. Warranties of habitability stipulate that a tenant's obligation to pay rent is contingent on the landlord maintaining the rental premises in good repair. The law has existed since the 1970s, but it continues to fail to provide a meaningful defense in eviction proceedings. As Nicole

Summers demonstrates, judges in eviction courts are too often nonplussed by the conditions of rental units—even when they have information about deplorable conditions right in front of them, and even when tenants have legal counsel.[15] Situating a building inspector in eviction court might be one way to increase the weight of evidence of bad housing conditions.

We could also institutionalize tenants' input concerning their housing. Perhaps tenant unions, along with government committees or community groups, could decide which issues landlords should fix, and what repairs—if any—they might consent to pay more rent to cover. Models already exist, and there is a long history of tenant unions. Contemporary tenant unions often prioritize preventing and protesting evictions, displacement, and gentrification. Adding another tool to their belt and campaigning for tenant control over repairs and housing conditions could help them in these fights. Indeed, tenant input into repairs could prevent landlords from using code enforcement as an excuse for dramatic upscaling that leads to displacement. Of course, tenants may not have the expertise to make decisions about some building codes, such as regulations about fire ratings, construction materials, and the importance of preemptive building maintenance. As such, a workable solution may be for tenant unions to consult building inspectors, and for municipal building departments to create a consultant position to institutionalize this practice. Under such a model, inspectors would have another recourse to assist tenants, which might change how they conceive of tenants and their position within the stacked deck.

Reshuffling the Deck for Homeowners

The stacked deck means that homeownership can be a burden. Properties deteriorate without maintenance and repairs. This has consequences beyond people being stuck living in dangerous or unpleasant conditions: properties depreciate, and unaddressed repairs can lead to liens, foreclosures, and displacement. State assistance is necessary to protect property owners without personal or familial wealth. The good news is that templates already exist. Most US cities have a program in place to assist low-income homeowners with home repairs.[16] The City of Chicago, for instance, offers assistance with emergency home repairs of up to $30,000. Since 1976, this program has fixed roofs and porches for eligible homeowners who apply through a lottery system.[17] Homeowners are eligible if their property is "habitable," owner-occupied, and "not at risk of foreclosure," and if they earn 80 percent or less of the area median income.[18] Conversations with recipients suggest that this program is effective at protecting them from unwanted moves.[19] However, there are limits to what repairs are covered, and as such the programs do not resolve all violations on record for a building. This means that property prices still depreciate, even if conditions improve for homeowners. Moreover, because of eligibility stipulations, this program leaves out those in the most precarious positions: homeowners whose homes are at risk of foreclosure and are not deemed habitable. The program also only assists those people lucky enough to have won its annual lottery.

Repair programs could also prevent building court from being a justice blocker. As others insist,[20] low-income defendants (property owners in this case) should be provided with free legal advice. But this is not enough. Homeowners also need resources and expertise to make repairs. Based on the premise that citizens have the right to a public defender if they cannot afford a lawyer, two Chicago architects recently outlined their vision for a public architect—free architectural guidance for property owners who need to fix violations that require an architect but cannot afford the professional fees architects charge.[21] Schemes like this—aimed at ensuring access to legal assistance and technical expertise—could also help to shift the focus of building court. Currently, the name of the game in building court is too often dismissal without ensuring that issues get resolved, which leads to deteriorating and sometimes dangerous conditions for property owners without resources, as well as for their tenants. Access to expertise and resources through government programs can ensure that building court is not a justice blocker, and it can give court actors—attorneys, judges, and inspectors—a recourse to assist people.

Reshuffling the deck necessitates that cities expand programs for home repairs and include those who are currently excluded. Cities could implement a system through which they allocated a certain amount of money to homeowners based on the assessed cost of necessary repairs, but one that also accounted for the income and wealth of a property owner. Such a system should mean that those most in need receive assistance with repairs. However, to be effective in preventing property depreciation, programs should expand what repair work is eligible, including masonry, carpentry, plumbing, and electrical work for example, in addition to emergency roof and porch repairs. Governments could focus repair assistance on property owners who are at risk of foreclosure, rather than leaving them out in the cold. Grants should take the place of current receivership programs, which—as public-private partnerships—are oriented toward profit and do little to reshuffle the deck. With financial assistance, code enforcement does not have to add to the headaches of small landlords, cloud property titles, or bankrupt struggling property owners.

Programs could also target neighborhoods from which homeowners are at risk of displacement. Again, models already exist. For example, Chicago recently passed the Woodlawn Housing Preservation Ordinance, which is aimed at preventing existing residents from losing their homes amid rapid change and upscaling. One section of the ordinance aims to provide home repairs to low-income homeowners in an effort to prevent displacement. Evaluation of such ordinances will be critical to determine the potential for reshuffling the deck through place-based initiatives. There are additional possibilities to target those most at risk. Rather than working with banks in questionable public-private partnerships, city governments could instead rely on their data. For example, banks, in their quest to save face amid their inequitable lending practices, could identify property owners for whom they underwrote risky mortgages. Cities could then offer grants to these property owners first.

Moving forward, the federal government could reimagine its role in the housing industry, perhaps by reallocating funds from the mortgage interest deduction program to an expanded home warranty program, which could offer a safety net for low-income homeowners. This would go a long way toward preventing newcomers and developers from profiting from dilapidation. Dilapidation also affords newcomers and external actors a sense of justification for the buildings they buy and the neighborhoods they move into because of the "improvements" they make to the built environments.[22] A lack of dilapidation could also nullify the need—either actual or rhetorical—for external investment, the arrival of newcomers, and the rhetoric of improvements.

Reshuffling the deck also necessitates dealing with the obstinacy of racial disparities in housing conditions. Making a dent in racial disparities in housing inequality requires explicit redistributive practices to compensate for the legacies of state-sanctioned discrimination and uneven expropriation of wealth in Black and Latinx communities that manifest in vastly disparate housing conditions between communities of color and their White counterparts.[23] Currently, a handful of city governments are beginning discussions about implementing reparations. Evanston, Illinois, for example, has approved the Restorative Housing Program, which will provide up to $25,000 to Black residents. Yet restrictions on how funds can be used mean that homeowners may end up using these funds to pay mortgage principal or interest, which would have the program lining the pockets of banks.[24] This is particularly problematic considering that banks are the source of so much racial inequity in housing. Yet, without some such schemes that take aim at racialized housing inequality, building code enforcement is likely to continue to burden minority homeowners. At the very least, governments should do more to regulate the real estate industry and property tax assessments to prevent undervaluing and overassessing property in communities of color.[25]

Continued conversations about reparations are also important because they have the potential to change how we think about government interventions. Reparations center interventions as compensation for past laws and policies, and thus cast state interventions as necessary steps in amending the government's role in housing inequality as a matter of justice rather than a handout.[26] Overall, reorganized, progressive state intervention in housing markets is necessary to reshuffle the racialized deck. This is also the best way to avoid "White blind spots" or other failures to recognize the extent to which the deck is stacked in favor of Whites. We must reshuffle the deck so that it is not stacked so firmly in favor of one population at the expense of another.

The end goal of state interventions should be to decouple homeownership from the promise of wealth and security. As Keeanga-Yamahtta Taylor recently argued, the housing industry is predicated on homeownership being a necessary form of security for US citizens because of the absence of social safety nets.[27] Governments need to reorient their approach to housing with two objectives in mind. First, we need to dissociate housing from the private sector.[28] This will require dismantling the instruments—such as mortgage securitization and real estate investment trusts—that allow extreme profit from housing. In place of these tools of financialization, the government should construct and manage publicly owned housing, commit to funding community-owned housing, and adopt policies that more effectively take control of empty buildings and vacant land.[29] There is growing evidence that we can approach property in ways that depart from dominant understandings of private ownership of property as a source of financial investment.[30] Reimagining our relationships to property goes hand in hand with the second goal: to provide a social safety net. Without free or affordable healthcare and education, quality of life depends on personal accumulation of wealth.[31] Americans need wealth, and homeownership is sold as the best way to get it. Governments must provide healthcare and education so that homeownership is not seen as the only path to health and opportunity, and less of our world rests on our access to homeownership.

Some of the policies and interventions I have reviewed may work better and be more immediately manageable than others, but it is critical we have as many ideas as possible to circumvent justice blockers and reshuffle the deck. Otherwise, we are ill-equipped to prevent the poor from staying poor and the rich from getting richer. If we fail to reshuffle the deck—for renters, homeowners, or anyone else—attempts to mete out justice will do little to challenge injustice. As this conclusion has laid out, without the guarantee of resource redistribution, we need a variety of plans and policies, because effective justice looks different for different people—for landlords versus tenants or low-income property owners, for example. Reshuffling the deck requires a foolproof system where differences in such things as resources and wealth are officially measured and acknowledged. Without such a system these differences will only be guessed at, as my book shows. People like frontline workers make assumptions about differences based on their perceptions of the stacked deck. But their assumptions are not always accurate, so we cannot rely on them to achieve justice. Instead we should harness this facet of frontline work to embrace the fact that differences and disparities are always going to inform decision-making. And with this information, we can ensure that justice is doled out in a way that evens out differences. Any effort at wholesale justice must be robust enough to account for the differences in contexts and beliefs that inform selective enforcement and implementation. If property owners had resources or safety nets, and if exploitation were prevented, discretionary decisions—which we can never avoid—would become less consequential. But it takes walking in the shoes of people on the ground to know this. They are the ones who work within the laws we pass, the policies and programs we roll out. If we listen and watch frontline workers, we can find out which proposed solutions are robust enough—which proposals are up to the job.

Zooming In to Zoom Out

Two things happened while I was writing this conclusion. A twelve-story oceanfront condo tower in Miami collapsed, allegedly because of costly structural defects, killing at least ninety-seven residents. Engineers had noted the structural issues, but nothing had been done. This disaster is the most recent in a line of building fires and collapses—from Grenfell Tower in London to the Hard Rock Hotel collapse in New Orleans—that should have been avoidable because people had prior knowledge that there were serious issues in the buildings. The Miami condo collapse stands out from these other incidents because its residents—owners of relatively pricey condos—were not who we typically think of as marginalized. The socioeconomic status of the residents further suggests that this tragedy could have been avoided. These are the kinds of people to whom society usually listens. So why was nothing done?

This question was the central focus of a second thing that happened when I was writing this conclusion. The *Chicago Tribune* published a front-page story claiming that at least sixty-one Chicagoans had "died in buildings the city knew were firetraps." The piece, "The Failures Before the Fires," alleged reckless indifference on the part of the City.[32] The article mentions inspections fifty-four times. Yet the voices of inspectors are notably absent. I shadowed inspectors for a relatively short portion of their twenty-plus-year careers, but I learned a lot from them that has helped me to assess ideas and proposals for changes in the city. Even though frontline workers like inspectors are the topic of articles like this and will be the ones implementing and enforcing new innovations, policy makers and housing organizations rarely—if ever—ask their opinions as interventions are being imagined and planned.

I suggest that if we listen to frontline workers, we can find answers to our questions about the avoidability of building-related disasters as well as persistent housing inequality. We all too often work backward and infer on-the-ground decisions of frontline workers from patterns in data sets, maps, or statistical models. In contrast, I stood in the shoes of frontline workers and listened carefully as they made the decisions that cumulatively create those data sets and maps. As a result, my book exposes an irony that we would never detect by only looking at spreadsheets, statistics, or maps of code violations. This irony is that we need to listen to individuals on the ground to see that the solutions to contesting inequality do not in fact lie at the level of individual decisions. Solutions lie instead in the structures and systems in which individuals operate. We have to zoom in to be able to accurately zoom out. This has been a goal of this book.

Zooming in reveals that inspectors know repairs do not get done—because of a lack of resources or because of exploitation and negligence. But they lack the power to do much about it. Even when they try to use their power to make a difference, they are not that effective. Problems cannot be solved at the level of code enforcement; rather, they must be addressed in the current housing landscape, where exploitation is unfettered and people, who are stuck between preserving their current wealth and preserving housing conditions, own buildings they cannot afford to maintain. Too often, even property owners with resources—who have been told that homeownership is the path to stability and wealth—just cannot afford the costs of participating in this American Dream as well as the expenses of maintenance. Zooming in to zoom out allows us to shift our focus to the precise reasons that inequality is stubborn and problems in cities persist.

Stacked decks—the layers of relational inequality that we all perceive—can motivate action, but in reality they are also made up of justice blockers. In short, stacked decks impede the actions they inspire. This paradoxical combination makes stacked decks persistent and explains why cities often remain the unequal way they are. It explains, for example, why it is that even though various people and entities—policy makers, lawyers, frontline workers, nonprofits, activists, and residents—keep trying to do something about affordable housing or racial disparities in the housing market, little ever seems to change. The problem is not that building code enforcement is lacking, but that, as inspectors realize, people either cannot afford repairs or are able to profit from not making repairs. The paradox also explains why some renters live in conditions that would shock a tenement reformer from the turn of the twentieth century. It is not just capitalism or financialization or systemic racism that explains the obduracy of material urban inequality, but rather the paradoxical relationship between individuals and the contexts in which they act. As this book has shown, unless we remove justice blockers, attempts to mete out justice frequently do little to make things better, and in fact often make things worse. It is only by recasting our understanding of the relationship between individuals and their contexts—between structure and agency—that we can clearly see why inequality is so hard to contest. But there is power in knowing this. We need to understand what we are up against in order to push it aside, see a way around it, take it apart, and make it better.

Acknowledgments

First, I want to acknowledge the time and patience of the Chicago building inspectors who allowed me a glimpse into their world. One inspector, and I think he knows who he is, deserves particular credit for his quest to show me all that inspectors do. Next, I thank my mentors for their support, their confidence in me, and all their early morning emails. Mary Pattillo's interest in the project and determination to sharpen ideas and sentences has made every page a hundred times better. I dreamed up the idea for this book in Wendy Griswold's class (and finished it in her home office in Hyde Park), and I thank her for her enthusiasm and straight talking. Lincoln Quillian was invaluable in helping me connect small details to big ideas.

At Northwestern University, I was fortunate to learn from and alongside: Niamba Baskerville, Claudio Benzecry, Laura Carrillo, Tony Chen, Jordan Conwell, Anya Degenshein, CC DuBois, Wendy Espeland, Gary Fine, Emily Handsman, Kevin Loughran, Marcel Knudsen, Alka Menon, Anna Michelson, Bob Nelson, Laura Beth Nielsen, Ann Orloff, Diego de los Rios, John Robinson, Brian Sargent, Wara Urwasi, Dhika Utama, Stefan Vogler, Celeste Watkins-Hayes, and Vincent Yung. A Northwestern Presidential Fellowship allowed me the time to focus on research during graduate school and I join a long list of people in thanking Northwestern's workshops, particularly the Culture Workshop. At Tulane University, I thank Kevin Gotham and Patrick Rafail for comments on a draft of my manuscript, and Katie Johnson and Laura McKinney for their input. I appreciate hallway conversations and general support from Mariana Craciun, Amalia Leguizamón, Camilo Leslie, Chris Oliver, Steve Ostertag, and Mimi Schippers. Thanks also to Rich Campanella for map making, Kelsey Ryland for help with geocoding, and Annika Bruno for research assistance. For obtaining housing market data, I thank Jay Cao, Anthony DeFusco, and Jessica Ruminski. I am especially grateful to administrative and library staff at Tulane and Northwestern.

I wrote a lot of this book during a week in a cabin on the Mississippi River, thanks to the Studio in the Woods writing retreat, and the book has benefited from the support of the Mark L. and Judith R. Schwartz Endowed Fund at Tulane University's School of Liberal Arts. A Carol Lavin Bernick Faculty Grant allowed me to receive feedback on the book from four scary-smart Tulane undergraduates: Annika Bruno, Aaliyah Butler, Jessica Galloway, and Ellie Goldzweig. And the Provost's Faculty Book Manuscript Workshop grant allowed me the pleasure of bringing Celeste Watkins-Hayes and Rachel Weber to New Orleans, where they provided the most engaged and useful feedback on my book manuscript imaginable. A fellowship from The American Bar Foundation/JPB Foundation Access to Justice Scholars Program and an ATLAS grant from the Louisiana Board of Regents afforded me time to focus on completing the manuscript.

The book, and my academic interests that prompted it, have also benefited from conversations with and support from friends, colleagues, and mentors at other institutions. Japonica Brown-Saracino gets the first mention for being my go-to person from the beginning. I also thank Bernadette Atuahene, Debbie Becher, Catherine Gillis, Claire Herbert, Katie Jensen, Terry McDonnell, Kelly Moore, Jon Norman, Jeffrey Parker, Anna Reosti, Lucy Rowland, Beryl Satter, Nicole Summers, Stacey Sutton, and Judy Wittner. I am very grateful to the book's reviewers and to Elizabeth Branch Dyson at the University of Chicago Press, who managed to be encouraging even in the subject lines of her emails.

I have undoubtedly benefited from my wonderful parents, who are tucked away in England. The same goes for my additional family in the US. Finally, I thank my favorite person and ancillary editor, Melissa, whose skills range from intolerance for jargon, thinking of synonyms, and listening to me talk about the book on dog walks, to drawing hearts next to sentences I've written that she likes. Our cats also helped me write a book by making sure I was awake by 6 a.m. every morning. Beyond editing and timekeeping, I thank Melissa and our little family for every moment of everything else.

APPENDIX A

Methodology

I began my quest by applying to be an intern at Chicago's Buildings Department. As an intern, I joined a handful of college students who performed a range of tasks—from photocopying and data entry to creating databases of elevators that were overdue for inspection with certificates that had expired. While my fellow interns volunteered their time for academic credit or work experience, I spent eight months at the Buildings Department trying to get to grips with the opaque world of building inspections. I wanted to know what inspectors said and did behind closed doors, what they did during inspections, how they made decisions, and what they thought of their jobs and the city in which they worked. Most of all, I wanted to know how code enforcement related to urban inequality—how their individual on-the-ground decisions amassed to mirror, reproduce, diverge from, or challenge patterns of economic and racial inequality in the city.

In the office, I was surrounded by inspectors as they got assignments, shared stories about their inspections, and cursed the office computer software they used to input inspection reports. I learned a lot during these months, but I do not refer to any of my office observations and interactions in this book. It did not feel right to me to record or repeat inspectors' conversations in the office because they were not aware that I had this book in mind. I also suspected that telling inspectors about my research interests—which I'd need to do for my university's IRB approval if for no other reason—before I had established rapport would have jeopardized my chances of getting access to interviews and ride-alongs down the line.

A month or so into my internship, I cautiously broached the subject of conducting some interviews with inspectors. Inspectors seemed nonplussed—perhaps slightly bemused—by my interest, but dutifully told me I needed to ask their supervisors for permission. I assured Buildings Department supervisors that no, I was not writing an exposé for the *Sun-Times*, and that I just wanted to know more about what inspectors did out in the field. I was relieved when I got word that I could conduct some interviews. A senior supervisor arranged half of my interviews and made available an empty room in the office building of the Buildings Department. I enrolled the other half of my participants during my internship, by approaching inspectors I met in the office. In total, I interviewed twenty building inspectors, all of whom were male. During my research, there were only five female inspectors at the Buildings Department (out of a total of almost two hundred). By and large, female inspectors were less forthcoming, less likely to engage in small talk in the office. I took this as an effort to keep their heads down in a male-dominated field. Indeed, one female inspector told me that she studied the building code in her spare time and felt she had "to work twice as hard" to show her expertise. Aware of this power imbalance and not wanting to add to their labor, I opted not to go to extra lengths to recruit female inspectors.

Therefore, my interviewees were all males, in their forties, fifties, and early sixties. Most of my interviewees had over ten years' experience on the job. Two were African American, three were Latinx and the remaining fifteen were White. This is, from my observations working in the Buildings Department, a fair representation of the racial composition of inspectors in Chicago.[1] The interviews, which I recorded unless inspectors preferred that I did not, lasted between forty-five and ninety minutes. I asked inspectors to describe their jobs, talk about their typical day, and discuss changes in the city and in their work. I enquired about the neighborhoods they frequented and the kinds of buildings, people, and issues they saw in those places. I asked how they made decisions, what they took into consideration during inspections, whether inspectors ever disagreed about what counts as a building violation, and how much supervision and oversight their work entailed.

I quickly noticed patterns. Inspectors usually began interviews by stressing that inspections are objective and impartial. For example, Keith, a White inspector with almost twenty years' experience, insisted:

> It's pretty easy ... there's no gray in the building code, it's basically black and white ... it's either broke or it's not broke, it's got to be fixed or it doesn't have to be fixed. It's pretty simple. You know, we take a snapshot of what the condition of that building was in that minute in that time and write it down on a piece

of paper ... it's not hard, you know. ... We're just building inspectors: it's broke, it's not broke, it needs to be fixed ... that's how easy this job is. You know, so if somebody says it's hard, it's not.

Keith's tone sounded somewhat defensive, as if he was trying to ward off my suspicions about corruption or bias. But his tone soon changed. In fact, sooner or later, every inspector I interviewed began to tell me—in one way or another—that violations can be subjective, that the code is malleable, and that they assess violations case by case. Thus, my quest became to determine what motivated inspectors and how they made their decisions.

Looking back, I think it helped that I had not yet been out on an inspection when I began interviews, because this meant I asked inspectors to explain what inspections were like. They gave me details and clarifications that they might not have offered if I had experienced an inspection for myself. Of course, "talk" can be "cheap"; what people say they do does not always align with what they actually do.[2] With this in mind, I was even more determined to see firsthand what inspectors did. I knew my interview data would be more valid if I saw inspectors doing what they told me they did, and I knew this was the best way for me to understand what motivated inspectors' decisions.

Navigating Access

I had hoped that my internship would enable me to go out on inspections. I knew I needed to see inspections in person if I wanted to really understand the work of building inspectors and the role of code enforcement in the city. I imagined I'd win inspectors and supervisors over by being a hardworking and reliable intern and by showing a genuine interest in their work. My plan did not go as I had hoped, and this was the source of much anxiety over many months. Like others who have studied city workers, I encountered a great deal of resistance.[3] Inspectors seemed willing to take me out on inspections, but their superiors in management positions were adamantly opposed to the idea. I believe their resistance was a combination of a few things. Mostly, I think the context—a city government office plagued by a reputation for corruption and headline-grabbing scandal—has made supervisors extraordinarily cautious about any scrutiny, particularly over discretionary decisions. I think they were concerned about my safety—and the department's ensuing liability—on inspections if something happened in a dangerous building. I also expect that, in part, I was just an unwelcome inconvenience and one more thing for them to juggle.

I tried to reason with supervisors, promising I just wanted to see what inspectors did and that I did not intend on writing an exposé. I made photocopies of a chapter of Robin Nagle's ethnography of New York City garbage workers and handed them out to provide an example, desperate to convince them that my study was non-threatening and might even paint building inspectors in a sorely needed positive light. I tried to flatter the supervisors, telling them I wanted to see the important work of code enforcement in action, to shed light on the misunderstood and underappreciated labor that goes into making our cities safe. Nothing worked. The answer was no every time, and I knew that I was becoming a source of annoyance, possibly threatening the access to building inspectors that I did have.

In some ways, my anxiety and uncertainty were productive, because they prompted me to seek out other sources of data (which I describe below) when it seemed increasingly unlikely that I would ever be allowed to go out on an inspection. After many conversations with my dissertation committee and other faculty members at Northwestern University, I eventually took action. I "went nuclear"—in the words of, and on the advice of, a mentor—which entailed my going directly to the Commissioner of the Buildings Department to seek permission. I knew this was risky because, if the Commissioner said no, I was out of options. I also knew that going over their heads would anger those in management positions who had consistently denied my requests. But it was my only option. To my delight, the Commissioner—who happened to have just finished reading Beryl Satter's *Family Properties*, which provides a history of Chicago's code enforcement—agreed. Finally, I was permitted to go out on inspections! As I had assumed, though, I had incensed those in mid-level management positions. I was curtly informed that I was no longer welcome as an intern. I felt that I had made enemies and that my hard-won access could be taken from me at any time. Getting the kind of access that I thought was necessary to write a book was like pulling teeth, and every interview and day of inspections felt like a big win on precarious borrowed time. As a result, however, I made every second count. I treated every ride-along and interview as if it were my last.

Ride-Alongs

My ride-alongs with inspectors took me all across the city, from places I knew well to areas I had never been. I saw block upon block of bungalows

METHODOLOGY

on the Southwest Side that I had only seen from a plane landing at Midway Airport, pockets of neighborhoods I never knew existed even though they were not far from my own, and the disinvested and partially desolate landscape of the far South Side. Initially, I felt that inspectors were trying to shield me from seeing some aspects of the city. One inspector admitted that he was wary of my going to high-poverty neighborhoods. As we were driving on the North Side one day, an inspector told me that the South Side is a "different world." "You don't need to see that," he continued, "you don't need to see the ghetto." The longer I spent in their company, however, the more inspectors seemed to want to show me a diversity of places. In the office and in building court, inspectors compared notes on the places they had taken me. One inspector told another: "You got to take her to Englewood!"—a highly stigmatized neighborhood on Chicago's Southwest Side. Another told me that to get the full experience, I needed to go on an inspection "where you feel itchy afterwards." It is telling that no one inquired as to whether I had visited a luxury downtown high rise, historic mansion, or any area without a bad reputation.

I observed over sixty inspections with six inspectors. Our days began with a pile of 311 requests. Some inspectors planned their itinerary for the day, others were more casual about the order of inspections. All followed the same pattern when they arrived at a building. Inspectors survey the outside of the building first, before ringing the bell or knocking on a door to seek access to the interior. Inspectors would tell me their first impressions of the building in these first few moments, often as we waited to see if there was anyone home to let us enter. In most cases, inspectors do not get access to the interior of buildings, and their inspections consist of examining exteriors. Inspectors talked me through what they noticed in real time, pointing to issues they could see and explaining their severity—or triviality—to me. As they wrote up their notes back in the car or as we were leaving buildings, they also routinely justified their decisions about which issues they planned to report and which buildings they would recommend for court.

Ride-alongs with inspectors did more than just allow me to observe inspections in action. Driving around the city—and often sitting in traffic—enabled long conversations about the places we were driving through as well as clarifying discussions about inspections we had just completed. After an inspection one afternoon, one inspector commented, "I'm explaining myself now to you, so I have to put it into words, but usually it's just natural." Asking inspectors to explain a decision that was second nature to them was a particularly fruitful technique, though I am aware that they may sometimes have found this tedious.

By and large, as with interviews, inspectors seemed happy to talk to me about their work. I think it helped that I am White, because it likely put most inspectors (who are also White) at ease. I also suspect it was beneficial that I am not from Chicago, or even the US, and I have a noticeable (British) accent. These things meant that inspectors assumed I knew next to nothing about the city and took time to explain aspects of Chicago to me. As we drove south on the expressway, for example, one inspector noted that the exit we were about to take was "for the South Side," and then proceeded to tell me about how the city is divided into North, South, and West Sides. These kinds of moments were useful because they shed light on how inspectors make sense of the city. I am sure other aspects of my own social location also shaped my research. Being a student aided my access, I expect, as inspectors did not take me too seriously (they asked repeatedly: "How's that paper coming along?" "Did you graduate yet?"). Being a White female in my earlyish thirties in a world predominantly occupied by White men over forty probably meant that I was privy to some comments, and not to others that I may have heard if I were male. I avoided mentioning that I was in a same-sex relationship, and I tried to cover my bases in terms of gender performance, dressing sensibly while trying to maintain some element of femininity. Dressing "sensibly," in my mind, meant I would be taken seriously and be allowed to enter treacherous buildings. Presenting some femininity, I assumed, meant I wouldn't raise any eyebrows and would be considered normal by inspectors. One thing I am sure of is that my love for Chicago sports helped establish rapport.

I took copious fieldnotes during and after ride-alongs, paying particular attention to what inspectors said, the characteristics of buildings, and how inspectors weighed decisions about what to do when they found building code violations. I audio-recorded inspections with some inspectors, but most preferred that I not do so. I also noted addresses of buildings we inspected, which allowed me to search Chicago's publicly available database of building violations and court actions so I could be sure of inspectors' eventual decisions and follow up on violations, conditions, and future sales.

Observing inspections also allowed me to confirm that inspectors acted in the ways they told me they did during interviews. For example, in interviews, inspectors insisted they went easy on family buildings on the South Side of the city. I found this surprising because of the wealth of research that attests to racial bias among frontline workers and disproportionate perceptions of physical disorder in minority communities. Similarly, another literature—which demonstrates the comfort with which Whites express racist views in front of other Whites[4]—led me to expect inspectors to use racial epithets in my company. I had strong assumptions that inspectors would mirror these other workers. However, out on inspections, I saw inspectors do what they had told me they do. I observed inspectors opting not to give out violations for minor issues in small rental properties in low-income neighborhoods of color. Still, I found it hard to shake my skepticism, especially because such observations contradicted my own assumptions and the findings of decades of extant research about frontline workers. I wondered if inspectors acted differently when I wasn't around. My cynicism spurred me to find additional data sources.

City-Wide Patterns

While my observations afford insight into inspectors' on-the-ground decisions, I use statistical analyses of multiple unique datasets to investigate the relationships between these decisions and trends in inequality at the city level. For this task, I built a dataset that merged three sources of data: 1) Chicago building violation records from 2006 to 2015 that I downloaded from the City of Chicago Data Portal; 2) a list of 311 requests in Chicago that concerned buildings from 2006 to 2015 that I accessed through a Freedom of Information Act request; and 3) American Community Survey 2008–2012 five-year summary data. I matched 311 request data and building violation data to their corresponding census block groups, using their points of latitude and longitude. I then matched block group–level aggregate data to American Community Survey data and used negative binomial regression models to investigate how the frequency of violations and outcomes of inspections vary according to census tract–level demographics such as socioeconomic status and race/ethnicity, as well as building characteristics such as tenancy, age, and type.[5] My findings show correlations between requests, violations, and demographic and building characteristics.

To a large extent, my quantitative findings suggest that my qualitative findings are representative of inspections over time and across the city. I found that income is positively correlated with violations at the block group level. A $1,000 increase in median household income per block group is associated with a 0.3 percent increase in violations.[6] But, as I have explained throughout this book, there was also a notable discrepancy between my fieldwork and my statistical analysis regarding the rate of building violations in communities of color. My quantitative analysis suggests that inspectors give out more violations in communities of color, which contradicts my qualitative observations of inspectors showing compassion to minority property owners. Was I being too sympathetic to inspectors? I asked myself this question over again as I was writing this book. Identifying this discrepancy prompted me to dig deeper into the way inspectors think about inequality and the stubborn persistence of racial disparities in housing. Indeed, my pursuit to explain this discrepancy prompted me to notice underlying and overlapping racialized aspects of inspectors' work and develop the concept of White blind spots. Incorporating these facets into my analysis gave me a better understanding of the obdurate character of the stacked deck.

I remain convinced that inspectors' sympathy for communities of color is genuine and exists alongside the overlapping frames they use to make sense of racial disparities in housing conditions. What's more, even if inspectors' actions toward homeowners of color stem from paranoia about being called out for racism (rather than a sincere sympathy), I still saw them opting not to cite buildings in communities of color. In sum, I believe that the discrepancy between my fieldwork and statistical analysis is attributed to racial disparities in housing conditions. However, an additional reason for the disparity may relate to the different units of analysis in my qualitative and quantitative research. While my statistical models use aggregate data on race at the block group level, my fieldwork revolves around individual property owners. As such, my quantitative research tells us more about how inspectors' decisions align with the racial makeup of residents of an area (who are not necessarily property owners, and indeed are less likely to be so in communities of color than in White neighborhoods) rather than of property owners. However, inspectors often do not meet property owners, and base their assumptions about property owners on their interpretation of neighborhood demographics anyway.

Having both quantitative and qualitative findings allowed triangulation—a methodological technique that analyzes multiple sources of data on the same process. The idea is that consistent findings across data sources increase validity and reliability. While, as I have just described, I used quantitative analysis to probe the validity of my qualitative findings, I also rely on my qualitative analysis to explain patterns captured in my quantitative analysis.

Legal and Regulatory Framework

As I grew to understand what inspectors did and how they made decisions, I found I wanted to know more about the ramifications of their decisions and the legal and regulatory framework in which their work is embedded. I started with the building code. Though I had become familiar with certain sections and specific ordinances during my fieldwork, I wanted to understand the full scope of violations that inspectors could record. I conducted content analysis of the Chicago building code by selecting each ordinance (ordinances correspond to violations) that inspectors had applied between 2006 and 2015 (n = 259), and categorized the ordinances in the following ways: 1) what building types the ordinance addresses (e.g., single-family homes or high rises); 2) what material features the ordinance addresses (e.g., chimney, walls, windows); 3) whether the ordinance is more qualitative or more quantitative (e.g., sanitation vs. height of a wall); 4) whether the ordinance is caused by a defect or dilapidation (e.g., height of wall vs. peeling paint); and 5) what justifies the ordinance (e.g., fire safety, sanitation). I categorized both the number of ordinances and the frequency with which inspectors record violations of these ordinances. This exercise allowed me to contextualize inspectors' discretion and understand the limits thereof.

Enforcing the building code often entails court cases. As such, I decided it was necessary to become familiar with the goings-on in Chicago's building court. Compared to getting access to go out on inspections, observations in building court were straightforward. I did not need permission to be in building court because it is considered a public place, and I could easily spend hours sitting in on hearings, taking notes, and getting acquainted with the fast pace of court cases about building violations. Chicago's building court groups building violation cases into eleven "court calls," ranging from specific calls focused only on heat complaints or exterior walls, to more general calls. I observed at least five cases in each call, though I focused my attention on the broad call for "occupied buildings with general code violations," which is divided into three courtrooms by geographic location (north, west, and south). Each courtroom has its own judge, city attorneys, and building inspector. Inspectors give testimony in these cases as expert witnesses. I divided my time between these courtrooms and observed over one hundred cases. In my fieldnotes, I paid particular attention to what inspectors said and how judges, attorneys, and property owners relied on or contested their testimony. These data informed my understanding of the consequences of building inspections and how inspectors assess these consequences. Building court judges seemed happy to talk to me, but I found city attorneys to be guarded and suspicious of my presence in court.

I also conducted a handful of observations in administrative hearings to get an idea of what happens when property owners must appear there. The lack of deliberation and the absence of building inspectors in this setting prompted me to focus my courtroom fieldwork on building court instead of administrative hearings. I did, however, access administrative hearing records for 2006–2015 through a Freedom of Information Act request. These data, which I geocoded, allowed me to map the geography of hearings as well as calculate the average fine. I also used online records—kept by the Illinois Circuit Court, the Cook County Assessor, and the Buildings Department—to look up building addresses and trace the outcomes of building inspections, court cases, and administrative hearings. These records reveal, for example, whether a property was sold and to whom, and the property value, as well as whether it went through foreclosure, got permits for a renovation, or was demolished. When records were not clear, I visited and photographed buildings to ascertain their state.

Violations and Price Changes

Code enforcement affects property beyond physical or legal changes, however, and I wanted to try to measure these effects. I used quantitative analyses to show the implications of building violations on property prices and rents. For this, I used the City of Chicago Data Portal's database of building violations from 2006 to 2015, the American Community Survey 2008–2012 five-year summary data, and databases of rental listings and property transactions in Chicago.[7] I merged these sources to create unique data sets and ran ordinary least squares regressions.[8] From the rental listings database, I extracted data on units that were rented (to different tenants) twice between 2010 and 2015. From the database of property transactions, I extracted data on properties that sold twice between 2010 and 2015 to calculate the difference in sale price for the same property.

Using data on properties at two points in time—with building violations between the two points—allowed me to capture the relationships between building violations and prices. Moreover, my analysis separates the effects of violations that have been resolved by property owners and landlords and those that have not. This method provides much stronger evidence of a causal effect than would be possible with cross-sectional data, which would limit my analysis to observing general correlations between property values, rents, and violations. Using repeat sales and rents is equivalent to using models with fixed effects for housing units, as it affords a high degree of internal validity by being able to hold property-level characteristics—that would significantly shape prices—constant, such as a building's age, size, or surrounding amenities.[9] To control for neighborhood differences in prices, I used American Community Survey data on the average percent change in home price/rent for each calendar year (2010–2015) for each of Chicago's census tracts. I assigned the appropriate percent change to each year gap for the properties in my data sets and used robust standard errors clustered at the tract level to account for correlation of observations within tracts. This analysis showed how sales prices and rents changed in tandem with code enforcement.

My analysis yielded four notable findings. First, a *rental unit* with *resolved* violations can be expected to *increase* in price, controlling for the average price change in the area, and renovations, among other factors. A 10 percent increase in resolved violations is associated with a 5.52 percent rise in rent, holding other variables constant. However, the effects of *unresolved* violations are not statistically significant regarding rent prices.[10] Third, in terms of *property price*, a property with *unresolved* building violations is associated with a *decrease* in value, controlling for the average price change in the area, difference in number of rooms, the number of violations at the first sale, and the difference in years between sales. Specifically, a 10 percent increase in unresolved violations corresponds to a 3.4 percent reduction in price increase on average.[11] Finally, the effects of *resolved* violations are not statistically significant in my property price analysis. Coupled with my fieldwork in building court, these statistical models gave me insight into what happens to properties after inspections and the aggregate long-term effects of individual on-the-ground decisions.

Neighborhood Case Study

I still felt there was more to understand about the connections between code enforcement and neighborhood context. The final piece of my data collection entailed a case study of one neighborhood in Chicago. I selected this neighborhood for three reasons. First, the neighborhood is diverse in terms of race, ethnicity, income, tenure, and building type and receives close to the median number of 311 requests about buildings. As testament to this heterogeneity, one inspector told me that—compared to other places that are relatively homogenous in terms of income, race, wealth, or building type—this neighborhood varied "building by building." Second, inspectors consistently record an average number of violations in the neighborhood. Similarly, 30 percent of building-related 311 requests were addressed within the Buildings Department's target time of twenty-one days, which is consistent with the Office of the Inspector General's 2017 audit, which found that the department met its goal for response times only 37 percent of the time.[12] Finally, I lived in this neighborhood at the time, which afforded me insight into neighborhood context and made fieldwork more practical.

Over a two-year period, I observed community meetings and events and conducted interviews with local politicians, residents, and landlords. Maybe it was because I was looking for it, but I was struck by the omnipresence of code enforcement. The alderman mentioned building inspections as a way to fight crime in the neighborhood, community members conflated dilapidated buildings with gang crime, and there seemed to be widespread agreement that building renovations would improve the neighborhood. I also wanted to use this neighborhood to develop a fine-grained analysis of code enforcement, from 311 request to inspection to citations, court cases, and property transfers. To this end, I matched one year of 311 requests to data on inspections and FOIA-requested data on administrative hearings and building court cases by building address for the neighborhood. This gave me a data set of the 451 buildings that had been inspected within one year to trace the kinds of requests that led to violations, and the kinds of violations that prompted court cases or other actions. I also walked the neighborhood to take notes on every building in the data set. I noted size, tenure, physical condition, age, and other features.

While all neighborhoods have their own contexts and characteristics, tracing the outcomes of inspections in one neighborhood sheds light on the connections between 311 requests, inspections, violations, and court cases. In this neighborhood, only 2 percent of 311 service requests over the course of one year resulted in an inspection but no recorded violations. Some of these calls were prompted by ice on the sidewalk (which is not in the purview of the Buildings Department), two were calls concerning a senior center, and others were about missing permits (which inspectors can look up online). Ten additional 311 requests were made about addresses that do not exist. In short, there are reasonable explanations for any 311 request that did not lead to an inspection.

Almost half of all inspections in this neighborhood resulted in a notice—i.e., at least one recorded violation but no further action. For example, a tenant's 311 call to report bedbugs in the building led to an inspection that uncovered nineteen violations but nothing else. The violations are listed online, although only three are listed as resolved. A little over a quarter of 311 requests in the neighborhood led to an administrative hearing. One resident reported that "tile is falling off the walls in the bathroom." The inspector was not able to enter the property when, after six months, he arrived for the inspection. Still, the property was cited for six exterior violations, including a dangerous porch. The property owner ended up in an administrative hearing and paid a $1,000 fine. Half as many inspections resulted in a building court case. An anonymous caller, for example, reported that "the condo association put in windows and wind sound is coming in, and they keep saying they are going to fix them, and they haven't after 10 months." After an inspection two weeks later, an inspector cited the building for twenty-one violations, ranging from work done without permits to sanitation issues in the building. As a result, the condo building faced a building court case. A further 10 percent of requests went to a hearing and then a court case. After multiple tenants called to report a leaking roof and mold in a big apartment building, two inspections turned up thirty violations. The property owner paid a $500 fine as a result of the administrative hearing and was also sent to building court.

To be sure, neighborhood contexts—from political pressures and housing market trends, to racial and socioeconomic demographics—will likely produce different patterns in other places. Yet having this information for one neighborhood provides a useful benchmark by means of which to compare code enforcement in other areas. The neighborhood case study also performed a kind of fact-checking function for me. Talking about code enforcement with landlords, residents, and elected officials confirmed many of my findings from fieldwork with inspectors. As such, it helped me reach saturation—the point at which I felt it was okay to leave the field and stop collecting data.

* * *

At the start of 2015, I'm not sure I knew building inspectors existed. By the end of the year, I was riding along on inspections, spending hours observing court cases about building code violations, working as an intern in Chicago's Buildings Department, and traipsing through my case study neighborhood in the snow to note down characteristics of buildings with violations. I planned to study one aspect of the city we did not know much about: code enforcement. As my research began, I quickly learned that code enforcement is integrated into the fabric of every corner of the city in a way I had not imagined. While not every person in a city has a landlord or a mortgage, or encounters the police, every building can be inspected. And, as I grew used to hearing, every building has violations. Building inspectors also go everywhere: to dilapidated rentals, luxury high rises, vacant properties, single-family bungalows, and everything in between. I tried to gather as much data from as many sources as possible. My overall aim was to be as sure as I could about my findings and the conclusions I was drawing. In doing so, I hope that this book is as accurate, relevant, and useful as possible.

APPENDIX B
Building Violation Counts

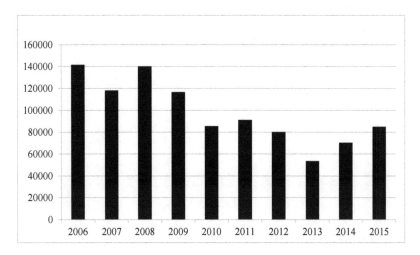

FIGURE 7. Annual counts of recorded building violations in Chicago, 2006–2015.

TABLE 2 **Building Violation Counts**

Community Area	2006	2007	2008	2009	2010	2011	2012	2013	2014	Total
01. Rogers Park	4701	2019	2328	2630	1436	1402	2677	1738	1451	20382
02. West Ridge	3985	1557	2551	2031	1529	1843	1762	1672	1591	18521
03. Uptown	3332	871	2907	1132	1295	970	1547	631	974	13659
04. Lincoln Square	1469	663	976	692	625	681	587	709	740	7142
05. North Center	735	559	833	602	592	456	1307	427	378	5889
06. Lake View	1645	1793	2613	2444	1697	1427	2265	1322	1908	17114
07. Lincoln Park	1256	901	1382	1436	694	876	1130	692	1101	9468
08. Near North Side	1396	825	3972	771	779	781	678	723	2834	12759
09. Edison Park	193	109	110	57	49	51	104	29	84	786
10. Norwood Park	401	301	347	315	163	246	341	349	698	3161
11. Jefferson Park	483	298	594	369	320	273	431	460	578	3806
12. Forest Glen	203	134	208	110	63	54	99	131	163	1165
13. North Park	442	437	1295	207	273	323	478	207	231	3893
14. Albany Park	2068	1219	2399	885	1250	1031	1467	1090	1356	12765
15. Portage Park	1963	1643	2100	1617	1117	977	1320	995	1477	13209
16. Irving Park	1405	1277	1905	1903	1180	1109	1559	1073	1666	13077
17. Dunning	639	572	671	707	414	427	552	658	544	5184
18. Montclare	276	298	237	229	183	261	261	273	165	2183
19. Belmont Cragin	2499	2549	2706	2259	1924	2749	2464	1764	1861	20775
20. Hermosa	1277	1197	998	1060	765	963	917	663	902	8742
21. Avondale	1200	1247	1744	1322	1057	1365	1763	971	1128	11797
22. Logan Square	3699	3220	4111	2670	2220	2370	2140	1827	1909	24166
23. Humboldt Park	4680	4425	5848	3583	3248	4124	4491	2610	2766	35775
24. West Town	3205	4795	5253	2850	4256	2653	2694	1685	1672	29063
25. Austin	8513	7848	8276	6108	5527	6951	5989	5112	5357	59681
26. West Garfield Park	2401	2396	3611	1862	1676	2409	2051	1528	1735	19669
27. East Garfield Park	2437	2548	2310	1838	1570	1928	1728	1243	1472	17074
28. Near West Side	1061	1294	1408	1805	2582	979	844	843	611	11427
29. North Lawndale	5517	5860	6178	3533	2790	3540	3287	2869	2976	36550
30. South Lawndale	3227	3140	3990	2587	2044	2740	2923	1492	1952	24095

31. Lower West Side	1427	1469	1773	3001	1319	1665	1538	759	1796	14747
32. Loop	368	937	1956	295	260	284	245	213	1307	5865
33. Near South Side	129	156	456	462	209	120	586	608	343	3069
34. Armour Square	256	270	396	161	180	296	255	660	184	2658
35. Douglas	312	458	338	340	314	350	275	577	301	3265
36. Oakland	64	83	195	205	97	114	63	70	62	953
37. Fuller Park	288	367	272	268	338	316	178	196	221	2444
38. Grand Boulevard	1443	1420	1740	2316	955	1080	1215	745	851	11765
39. Kenwood	329	373	995	662	457	414	422	233	544	4429
40. Washington Park	2532	2047	1340	1671	913	1001	1091	653	566	11814
41. Hyde Park	1055	442	745	806	429	535	508	166	445	5131
42. Woodlawn	2840	2030	1630	3499	1622	1920	1792	1266	1328	17927
43. South Shore	4230	4626	4714	7032	3362	4594	2825	2151	2485	36019
44. Chatham	2256	2282	2263	3234	1619	1754	1696	1390	1779	18273
45. Avalon Park	529	603	434	434	331	511	312	339	307	3800
46. South Chicago	4572	3278	2876	3951	2202	3621	2223	1629	1870	26222
47. Burnside	328	279	172	271	156	114	93	237	182	1832
48. Calumet Heights	507	664	436	421	363	328	277	216	291	3503
49. Roseland	4231	2715	2719	3869	2855	2948	2508	2220	2862	26927
50. Pullman	432	512	322	349	408	434	204	219	283	3163
51. South Deering	472	454	464	556	702	728	406	214	313	4309
52. East Side	566	837	655	583	834	862	355	204	385	5281
53. West Pullman	1939	1798	1804	1909	2161	2695	1811	1745	1818	17680
54. Riverdale	195	110	61	132	63	153	134	112	110	1070
55. Hegewisch	205	363	228	390	190	395	205	103	151	2230
56. Garfield Ridge	365	614	571	409	462	472	378	308	256	3855
57. Archer Heights	269	332	348	211	312	311	277	245	192	2497
58. Brighton Park	1677	1608	1900	1355	1098	2000	1316	981	686	12621
59. McKinley Park	423	888	786	807	512	595	575	661	334	5581
60. Bridgeport	1160	1527	2099	1363	1245	1301	1523	1813	810	12841
61. New City	3404	3331	3336	3316	2470	2714	2840	2802	2422	26635
62. West Elsdon	305	363	300	182	329	425	172	96	172	2344
63. Gage Park	1324	1444	1172	975	1318	1723	959	647	691	10253

continues

TABLE 2 *(continued)*

Community Area	2006	2007	2008	2009	2010	2011	2012	2013	2014	Total
64. Clearing	433	372	409	343	446	629	280	180	227	3319
65. West Lawn	471	597	484	380	848	1194	328	280	332	4914
66. Chicago Lawn	4358	3631	4022	3232	3082	4409	2114	1843	1973	28664
67. West Englewood	5607	4377	4760	4195	2968	3895	3674	3550	3311	36337
68. Englewood	5620	5520	5691	4643	3096	3998	3491	3556	3055	38670
69. Greater Grand Crossing	3696	3844	3869	3745	1845	2519	2372	1897	1906	25693
70. Ashburn	473	570	592	555	665	751	464	216	459	4745
71. Auburn Gresham	3825	3536	4093	3671	2615	3417	2307	1666	2216	27346
72. Beverly	288	384	272	414	438	341	185	83	229	2634
73. Washington Heights	933	1006	1219	1065	1058	1178	732	641	739	8571
74. Mount Greenwood	271	183	245	223	208	287	213	122	176	1928
75. Morgan Park	830	599	781	787	736	1156	537	402	496	6324
76. O'Hare	123	127	137	82	98	75	220	252	257	1371
77. Edgewater	2781	1087	1994	1302	1167	999	1284	963	1238	12815

APPENDIX C

Map of Strategic Task Force Inspections

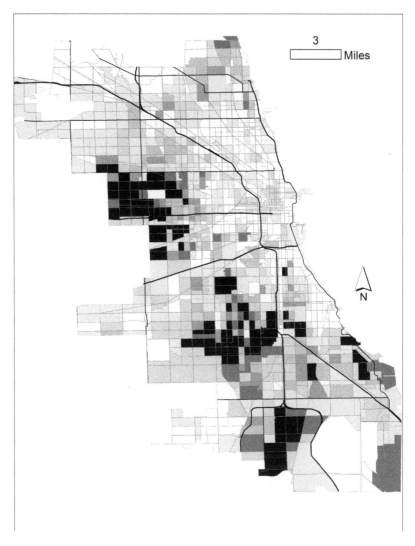

FIGURE 8. Strategic task force inspections in Chicago, 2006–2015. Darkest areas are well above mean, lightest areas are well below mean. Data are shown by census tract. Map by Richard Campanella.

Notes

Introduction

1. Accounts are pieced together using records of 311 service requests and building violation data.
2. Goffman (1974, 21).
3. Herbert (1996); Van Maanen (1978).
4. Desmond and Western (2018); Dwyer (2018); Garrido (2019); Vargas (2016).
5. Nor is the deck stacked by the "invisible elbow" that Tilly (1996) theorizes to explain happenstance consequences.
6. As Eubanks (2018, 200) argues, we need to build alternative systems, "on purpose, brick by brick and byte by byte."
7. Purifoye (2017).
8. Shedd (2015).
9. Nellis (2016). Also see Bobo and Thompson (2006); Pattillo, Western, and Weiman (2004); Van Cleve (2016).
10. Leicht (2008).
11. Sneath (2020).
12. Atuahene (2018); Center for Municipal Finance (2021).
13. Building on the work of theorists such as Marx and Tilley, scholars have recently renewed interest in the relational quality of inequality—see Desmond and Western (2018); Dwyer (2018); Taylor (2019); Tomaskovic-Devey and Avent-Holt (2019).
14. Desmond and Western (2018, 310).
15. Desmond (2016); Sassen (2014); Stuart (2016).
16. Taylor (2019, 37).
17. Rothstein (2017).
18. Solnit (2021).
19. Pattillo (2021). Pattillo overviews research by English et al. (2014); Okulicz-Kozaryn (2019); Pickett and Wilkinson (2008); Vogt Yuan (2007); and White and Lawrence (2019).

20. Pattillo draws on research including Blake (2018); Bennett and Lutz (2009); Kim and Conrad (2006); Merolla (2013); Mykerezi and Mills (2008); Owens et al. (2012); Sharkey (2013); and Wilson (2007).

21. Hunter et al. (2016) chart urban Black Americans' creation of sites of endurance, belonging, and resistance despite external assaults on Black spaces.

22. See Hollander and Einwohner (2004).

23. E.g., Beckett and Herbert (2008); Epp, Maynard-Moody, and Haider-Markel (2014); Harcourt (2001); Herbert (1996); Katz (2013); Quadagno (1994); Sampson (2012); Sampson and Raudenbush (2004); Soss et al. (2009); Wacquant (2009).

24. In this way, stacked decks act as "injustice frames" that help identify victims and perpetrators (Gamson, Fireman, and Rytina 1986).

25. On whether inequality is objective, see Bottero (2019).

26. For work on class, see for example Bourdieu (2013). For work on dispositions, race, and ethnicity, see Banks (2010); Rodriquez (2006); and Shively (1992). For gendered differences in dispositions, see Christin (2012) and Lizardo (2006).

27. Professional or political clout is tied to social location but not reducible to it.

28. Lewis and Diamond (2015).

29. Bartram (2021).

30. Runciman (1966).

31. Shedd (2015) shows Black adolescents in Chicago forming ideas about inequity through interactions with a racially stratified city. Minkoff and Lyons (2019) demonstrate that residents of economically diverse neighborhoods are more likely to perceive income gaps as problematic than are people who live in more homogenous neighborhoods.

32. Scott (1985).

33. Bottero (2019) argues that when we try to understand or measure people's attitudes to inequality, we are really capturing their sense of inequity—whether people think unequal situations are fairly generated. This explains seemingly inconsistent survey results that people in more unequal countries express higher tolerance of inequality.

34. Brown-Saracino (2009).

35. Boudon (1982); De Zwart (2015); Merton (1936); Portes (2000).

36. Watkins-Hayes (2009a).

37. Lamont, Beljean, and Clair (2014) argue that cultural processes are largely uncertain and open-ended, but channeled and magnified by institutional contexts.

38. Sullivan (2017).

39. Summers (2020).

40. Mijs (2018); Minkoff and Lyons (2019); Runciman (1966); Shedd (2015).

41. Lipsky (2010); Soss et al. (2009); Watkins-Hayes (2009a); Zacka (2017).

42. Lara-Millan (2017); Prottas (1979); Soss et al. (2009).

43. E.g., Krinsky and Simonet (2017); Stuart (2016); Watkins-Hayes (2009a).

44. Klepeis et al. (2001).

45. Between 2013 and 2017, house fires caused 2,620 deaths, 11,220 injuries, and $6.9 billion in direct property damage in the US (Ahrens 2019).

46. E.g., Schram (2000); Stuart (2016); Wacquant (2009).

47. E.g., Fording, Soss, and Schram (2011).

48. Hong (2016); Lewis (2000); Sun and Payne (2004). For variation on these findings, see Singla, Kirschner, and Stone (2020); Watkins-Hayes (2011); Wilkins and Williams (2008).

49. Inspectors are one of the "many hands of the state." Morgan and Orloff (2017) use this term to move away from conceptions of state as unitary. See also Valverde (2011), who argues that urban governance is flexible and contradictory.

50. Watkins-Hayes (2009b) theorizes that the combination of personal history, institutional politics, and environmental phenomena shape how government employees do their jobs.

51. Young (2011).

Chapter One

1. On uneven urban development and disproportionate neighborhood investment, see Smith (1991).

2. Data on median household income is from the 2010 Census. Data on White population is from the American Community Survey 2008–2012 five-year summary. To create the maps, we used Arc GIS Jenk's algorithm to find natural breaks in the histogram for each variable. We then manually adjusted for clean increments for cartographic neatness.

3. See Keating (2008) for details on the construction of Chicago's sides.

4. See Bey (2019); Black (2019); Moore (2016).

5. Pattillo (2013).

6. Moore (2016).

7. See Black (2019); Hirsch (2009).

8. Archives also show that covenants were enforced (Plotkin 2001).

9. Hirsch (2009); Satter (2009).

10. Michney and Winling (2020).

11. Satter (2009).

12. A study of Black property purchases from 1953 to 1961 on four blocks in this Southwest Side neighborhood found that installment contracts were the principal means of purchase by Black homebuyers (88 percent). On average, contract purchases cost buyers $587 (in 2019 dollars) more per month than conventional mortgages (George et al. 2019).

13. Satter (2009).

14. Hirsch (2009); Hunt (2009); Satter (2009); Vale (2013).

15. Taylor (2019).

16. Taylor (2019).
17. On property tax liens and racialized systems of taxation, see Atuahene (2018); Harris (2004); Hendon (1968); Henricks and Seamster (2017); Kahrl (2015, 2016); Passell (2021).
18. Wilson (1996).
19. See Aalbers (2019); Christophers (2020); Fernandez and Aalbers (2016); Gotham (2009); Immergluck and Law (2014); and Rolnik (2013). To make housing an increasingly transferable asset, the federal government, investment banks, and finance capitalists created instruments like mortgage securitization and real estate investment trusts, along with new tools and conventions for advanced real estate valuation, categorization, and rating.
20. Banks then combine thousands of loans (including car loans, credit card debt, and student loans) into derivatives to sell to investors.
21. Financialization increases indebtedness across the board, incorporating previously excluded groups into financial entanglements (Aalbers 2019).
22. Foreclosure crises have occurred periodically after finance-driven bubbles. See Immergluck and Law (2014).
23. Hall, Crowder, and Spring (2015); Thomas et al. (2017); Rugh and Massey (2010); Rugh, Albright, and Massey (2015).
24. Gamino (2015).
25. Some Southwest Side neighborhoods averaged 546 foreclosure filings per year from 2007 to 2012. This figure is well over double the citywide average of 260 per year (Institute for Housing Studies 2020).
26. Institute for Research on Race & Public Policy (2016).
27. All but one of the eighteen Chicago community areas with the highest foreclosure rates are predominantly Black or Latinx—Institute for Research on Race & Public Policy (2016); Thomas et al. (2017).
28. Foreclosed properties are nonperforming assets, which cost banks by hindering their capacity to borrow money and decreasing their credit rating.
29. Institute for Research on Race & Public Policy (2016); Cowan and Aumiller (2014).
30. Of all foreclosures filed between 2008 and 2010 in Cook County, approximately 8.7 percent (more than 11,700) were zombie properties (Woodstock Institute 2014).
31. Parker (2018).
32. Hackworth (2014).
33. Smith (1987).
34. Fields (2018); Fields and Uffer (2016); Immergluck and Law (2014).
35. On the role of finance in Chicago's downtown built environment, see Weber (2015).
36. Christophers (2020).
37. Chicago's segregation makes it likely than many property owners in communities of color are people of color, and property owners in White communities

are White. For details about race and property taxes, see Kahrl (2016); Massey and Denton (1993).

38. Howell and Korver-Glen (2018).

39. See Krysan and Crowder (2017) for research on residential decisions and segregation.

40. Besbris and Faber (2017); Korver-Glen (2018, 2021).

41. These data are for the years 2012 to 2018 (Lutton 2020).

42. Institute for Research on Race & Public Policy (2016).

43. Adler and Rehkopf (2008); Conley (2010); Joint Center for Housing Studies (2017); Sampson and Winter (2016).

44. Fishman (1987); Jackson (1987); Massey and Denton (1993).

45. Choi et al. (2019).

46. Desmond (2017); McCabe (2018).

47. McCabe (2016).

48. Flint (1977); Rothstein (2017).

49. Rothstein (2017).

50. Hertz (2018, 13).

51. On neighborhood branding, see Mele (2000).

52. Korver-Glenn (2021).

53. Halle (1993); Kefalas (2003).

54. Harris (2013, 3).

55. Brown-Saracino (2009); Summers (2019); Zukin (1987).

56. Rotella (2019).

57. Keating (2008).

58. Moore (2016); Rotella (2019).

59. Zukin (1987).

60. Jacobs (1961); Zukin (1998).

61. Rotella (2019).

62. Rotella (2019).

63. Ibata (1982).

64. Straw buyers are people who make a purchase on behalf of someone who is not legally allowed to make the purchase.

65. McKenzie (2019).

66. United States Department of Justice (2012).

67. The Illinois Distressed Condominium Property Act addresses disrepair due to fraud, unexpected expenses, or incidents when owners vacate and associations are unable to cover costs of upkeep. Declaring condos "distressed" empowers the court to deconvert the building to a single owner (Illinois General Assembly 2010).

68. This is limited receivership. General receivership is when a judge gives an agency the title outright. Receivership is distinct from eminent domain in that the City does not take the title to a property, but rather transfers it to another entity.

69. JPMorgan Chase, Chicago Community Development Financial Institutions, the US Department of the Treasury, and the Federal Home Loan Bank of Chicago

fund Chicago's largest receiver agency—Community Initiatives, Inc. (CII). CII "over[saw] the rehabilitation of 500 buildings with 9,800 units; purchased and transferred to responsible owners 245 buildings with 3,802 units of affordable rental housing; and acquired 289 condo units, filed condo deconversion orders and transferred 57 buildings with 581 units to new owners" between 2003 and 2015 (Community Investment Corporation 2020).

70. Community Investment Corporation (2018).
71. Rotella (2019).
72. Semuels (2019).
73. "Corporate landlord" is an umbrella term for real estate investment trusts, private equity funds, financial asset management firms, and other investment vehicles that began aggressively acquiring property in the mid-2000s—August and Walks (2018); Fields (2018); Fields and Uffer (2016); Teresa (2016).
74. US Census Bureau (2018).
75. See Gabbe (2015); Christensen et al. (2019).
76. August and Walks (2018); Fields and Uffer (2016); Teresa (2016).
77. Mallach (2014).
78. Immergluck and Law (2014); Mele (2000).
79. Garboden and Newman (2012); Satter (2009).
80. Klein (1964); Stegman (1967); Sternlieb (1966).
81. These strategies are facilitated by a trio of government actions: legislation for new vehicles of financial investment in real estate (such as mortgage securitization—see Gotham 2009), withdrawing funds from social housing (see August and Walks 2018), and the creation of LLCs that allow landlords to avoid risk (see Travis 2019).
82. Blessing (2016); Fields and Uffer (2016); Robinson (2016); Tapp and Kay (2019).
83. Rosen (2020).
84. Theodore and Martin (2007).
85. Centro Autónomo de Albany Park (2020).
86. Keating (2008).
87. On this divergence, see Brown-Saracino (2017).
88. Huq and Harwood (2019).
89. On policing and gentrification, see Beck (2020); Laniyonu (2018).
90. STF inspections are part of the Municipal Drug and Gang House Enforcement Program—Chicago Police Department (2012).
91. Doering (2016); Skogan (2006).
92. Hernandez (2020).
93. Metropolitan Tenants Organization (2010).
94. Centro Autónomo (2016).
95. Tenants can legally withhold either half of their rent payment or a maximum of $500.
96. For example, the Metropolitan Tenants Organization and the Lawyers' Committee for Better Housing.

NOTES TO PAGES 44–54   191

97. As elsewhere, landlords in Chicago organize and lobby against increased tenant protections such as rent control, further stacking the deck against tenants.

98. Brown-Saracino (2009); Zukin (1987).

99. Institute for Housing Studies (2020).

100. Centro Autónomo (2016); Progress Illinois (2015). However, it can be difficult to count how many people are displaced—Newman and Wyly (2006).

101. Desmond (2016); Purser (2016).

102. Lawyers' Committee for Better Housing (2019).

103. Woldoff, Morrison, and Glass (2016); Zuk and Chapple (2016).

104. Nearly twice as many Black residents rent as own their homes, and Latinx residents are also predominantly renters (Institute for Research on Race & Public Policy 2016).

105. Desmond, Gershenson, and Kiviat (2015).

106. City of Chicago (2019b).

107. Centro Autónomo (2016).

108. Christophers (2019).

Chapter Two

1. To protect inspectors' identities, I use pseudonyms and I do not name neighborhoods when recounting inspections.

2. Throughout this book, I use data from the ten-year span of 2006–2015. 2006 is the first year for which publicly available data on buildings violations exist.

3. I geocoded ten years' worth of FOIA-requested 311 request data and matched it to census tract data, to normalize the number of requests by number of housing units per census tract. Akin to standardization, normalization is the process of dividing one value by another to minimize differences in values based on the size of an area or the number of features in an area. The dark lines on the map represent major thoroughfares.

4. Between 2006 and 2015, inspectors recorded 883,183 request-prompted residential violations of 256 different code ordinances. Appendix B breaks down building violations by year and at the neighborhood level. The range of shades is narrower in this map than in the first one because the maps depict standard deviations from the mean. ArcMap calculates the mean and standard deviation. Class breaks are created with equal value ranges that are a proportion of the standard deviation—usually at intervals of 1, ½, ⅓, or ¼ standard deviations using mean values and the standard deviations from the mean. The maps in this chapter use the ½ interval, which produced seven categories, just enough to express in grayscale. Dividing up the data based on the inherent structure of their distribution means that the method is consistent across maps, if not in the legends.

5. The Voorhees Center (2001).

6. Watkins-Hayes (2009a).

7. In the absence official data, demographic information is based on my fieldwork observations.

8. The demographics of Chicago's building inspectors are representative of inspectors across the US, where 85 percent of inspectors are over forty-five years of age; most began their careers in construction trades and do not have a bachelor's degree (International Code Council 2014). There are no data on race or gender. However, since most inspectors began their careers in construction, a majority is likely White and male—see Goldberg and Griffey (2010).

9. Connell (2006); Collinson (1992); Lamont (2000).

10. Halle (1984); Lamont (2000); Kefalas (2003).

11. Kefalas (2003).

12. Other studies of frontline workers also highlight consistency across racial groups. Lamont (2000) finds Black and White working men value very similar moral dimensions (such as hard work, straightforwardness, sincerity, and consistency), but also slight differences. She finds that Black working-class men are more likely to link poverty to structural (rather than individual) explanations and are more critical of exploitation. The Black working men in her study also buy in more wholeheartedly to the American Dream than do White men. Lamont suggests that White working-class men are more individualistic due to the demise of unions. However, the prevalence of unions for both White and Black building inspectors in Chicago, as well as entrenched political patronage, may diminish this difference somewhat.

13. The starting salary for building inspectors is approximately $70,000, which is comparable to that of the Chicago Police Department (City of Chicago 2017b). Despite their high salaries, inspectors are steadfastly working-class. The type of occupation, lack of workplace authority (inspectors have discretion and clout, but lack authority), and level of education make inspectors comparable to the men Lamont (2000) describes as working-class. Inspectors also live in working-class neighborhoods. They may have the financial resources to move to suburbia (but cannot because their jobs mandate their city residency), yet not enough money to move to higher-income urban neighborhoods. Their colleagues, families, and residential environments are all working-class, which reinforces their class disposition.

14. Most Americans claim to value fairness and hard work and have contempt for unfair profit (McCall 2013).

15. Katz (2013); Maynard-Moody, Musheno, and Musheno (2003); Soss et al. (2009).

16. Herbert (1996); Portillo and Rudes (2014); Stuart (2016); Wellman (1993).

17. Seim (2020).

18. Soss et al. (2009, 11).

19. Inspectors mentioned Chicago Police officer Jason Van Dyke, who killed Laquan McDonald, an unarmed Black teenager, by shooting him sixteen times. One inspector expressed outrage at the murder, yet also confessed that he was glad inspectors did not carry guns and so were unlikely to "end up like Van Dyke."

Specifically, he referenced being alarmed by noises in empty buildings that might prompt inspectors to act in haste and shoot if they were armed. Racism in Chicago government is not restricted to the police department. In 2017, employees in Chicago's Water Department were found to have sent racist and offensive emails.

20. A fear of being watched pervades the Department of Buildings. I quickly learned that employees prefer to communicate in person because, as government workers, their emails "can be FOIA-ed," in the words of one inspector. The FBI has also tapped departmental phonelines during corruption investigations.

21. To the best of my knowledge, supervisors endorse inspectors' decisions. However, the absence of formal rules directing inspectors' discretion means that those in managerial roles are shielded from potential criticism for any decisions that inspectors make. While there is little in the way of immediate oversight, inspectors are always aware of the possibility that their reports might be checked and want to avoid being called "into the office."

22. Halle (1984).
23. Tonkiss (2005, 60).
24. Bartram (2021).
25. Halle (1984); Kefalas (2003); Lamont (2000).
26. Harris (2013); Kefalas (2003).
27. Bourdieu (2013) refers to this as symbolic domination.
28. Oliver and Shapiro (1995); Keister and Moller (2000).
29. Bartram (2021).
30. For a theory of how consensus emerges in this way, see Rawlings and Childress (2019).
31. According to the City of Chicago's budget overview, the annual budget for the Buildings Department was between $34 million and $37 million during my fieldwork (0.39 percent of the city's budget on average).
32. Chicago makes more of its revenue from fines and forfeitures than other large US cities. In 2017, for example, Chicago raised 4.9 percent of its revenue from fines and forfeitures, compared to the 1.4 percent average across the largest US cities. This amounts to $321 million ($118 per capita) in fines and forfeitures. See Calame and Boddupalli (2020).
33. City of Chicago (2018).
34. Valverde (2011).
35. Building codes are a kind of regulation of private property that is tolerated in the US because they are motivated by a desire to maintain the right to profit from private property (Valverde 2011).
36. For a history of Chicago's building code, see Satter (2009); Seligman (2005).
37. Unlike most municipalities in the US, Chicago uses its own building code and has not adopted the International Building Code (IBC). Chicago is less of a special case than it appears, however, because many municipalities retain large segments of their city-specific building codes in addition to the IBC.

38. Department of Buildings (2017).

39. Based on content analysis of the Chicago building code—see appendix A.

40. City building inspectors only inspect the exterior of buildings occupied by Housing Choice Voucher recipients, but they do inspect Chicago Housing Authority (CHA) buildings. However, the Buildings Department cannot prosecute other city or state entities (such as the CHA), thus citations neither lead to court cases nor mandate enforcement.

41. At the time of my research, the Department employed fifty fewer inspectors than it had ten years previously (Chicago Office of Inspector General 2018).

42. This Bureau stems from the Illinois Urban Community Conservation Act of 1953, aimed at preventing blight and conserving areas at risk from decay—Satter (2009).

43. The Buildings Department does not track positions by bureau (Chicago Office of Inspector General 2018, 20).

44. The ordinances to which violations correspond apply practically equally to buildings, regardless of construction material, size, or use. Less than 1 percent of ordinances distinguish between frame and brick buildings, for example. And while 7 percent of ordinances do not pertain to single-family homes, these ordinances make up only just over 1 percent of all ordinances inspectors used between 2006 and 2015. Moreover, 68 percent of ordinances apply to physical aspects found in all buildings (e.g., windows, doors, walls, and floors vs. chimneys, stairs, and yards). Furthermore, the existence of some comprehensive and nonspecific ordinances means that inspectors can use them to cite items that are not named specifically in the code. For example, while there is an ordinance for carpets, there is not one for hardwood floors. However, an inspector can still record an issue with a hardwood floor under the general ordinance for floors.

45. Residential inspectors also conduct annual inspections of buildings over three stories. However, the Buildings Department prioritizes 311 requests and only completes a small portion of annual inspections each year. Between 2006 and 2015, over 73 percent of all building violations were prompted by a 311 request.

46. Of the twenty-five most populated US cities, only three do not have 311 systems in place.

47. City of Chicago (2019a).

48. This is commonly known among inspectors and corroborated by my analysis of 311 requests (see appendix A).

49. Chicago Office of the Inspector General (2018).

50. See appendix A for a discussion of the very few 311 requests that do not result in an inspection.

51. Satter (2009); Seligman (2005).

52. Federal Bureau of Investigation (2009).

53. Based on my analysis of cases cited in Gradel and Simpson (2011).

54. This is accurate for the subset of inspections that result in violations, which is nearly all inspections.

55. Twenty-one of the twenty-five most populous US cities list building violation records online. In Chicago, many of these records are updated daily, as departments upload their data as it comes in.

56. Fines ranged from $200 to $50,000. The latter is the maximum fine.

57. Not all cities have two different court systems for building violations. In some municipalities, a centralized building court covers all housing issues. I use the term "building court" to distinguish between the courtrooms that cover building violations and those that deal with other housing issues, such as evictions. However, during my fieldwork, inspectors and court actors used the term "housing court."

58. Unlike regular residential inspections, court inspectors schedule inspections.

59. Inspectors refer to 311 requests as complaints. I opt not to follow their lead, however, because of the negative connotation of the word "complaint."

60. Mallach (2018, 3).

61. Civic Consulting Alliance Board (2018); Hinz (2018).

62. See Sanchez and Kambhampati (2018); Singla, Kirschner, and Stone (2020).

63. Eubanks (2018).

64. Novak (1996); Valverde (2012).

65. Fishman (1987); Hirsch (2009); Jackson (1987); Massey and Denton (1993); Rothstein (2017).

66. For exceptions, see Proudfoot and McCann (2008); Sommers (2016); and Valverde (2012).

Chapter Three

1. Keating (2008).

2. We could not inspect this tenant's unit because she was under eighteen years of age and no adult was present.

3. Bartram (2019a).

4. Keating (2008).

5. Sampson (2012); Sampson and Raudenbush (2004).

6. Katz (2013); Soss et al. (2009); Quadagno (1994).

7. On distinctions between landlords, see Balzarini (2017); Garboden and Rosen (2018); Shiffer-Sebba (2020); and Sternlieb (1966).

8. Keating (2008).

9. Plumer, Popovich and Palmer (2020).

10. Desmond (2016).

11. For an overview of discrimination in the rental market, see Krysan and Crowder (2017); for studies that show how evictions limit future rental options and reproduce inequality, see Desmond, Gershenson, and Kiviat (2015); Purser (2016); and Reosti (2021). Desmond, Gershenson, and Kiviat (2015) also demonstrate the consequences of eviction for households with children.

12. The neighborhood's diversity is a result of a series of policy decisions on the part of local elected officials to curtail affordable housing and appease White residents (Berrey 2015).

13. Bartram (2021).

14. On micro-units, see Christensen et al. (2019); Gabbe (2015).

15. For a geography of building size, see Institute for Housing Studies (2013).

16. Bartram (2019a).

17. However, single-family homes comprise a relatively small percentage of Chicago's rental stock compared to other cities (Kurth 2012).

18. Small-time landlords are also more likely to be both owners and landlords, whereas the owners of big buildings contract out management. Inspectors do not seem to make this distinction. While they usually refer to landlords rather than owners of rental buildings, I believe this is just to help distinguish them from owner-occupiers who do not rent out their buildings.

19. Institute for Housing Studies (2020).

20. See Christophers, who claims that "income derives simultaneously from control of an asset and from the work involved.... Few, if any, assets auto-generate income. They ... need to be put to work" (2020, 6).

21. On investment in property and the evaluative schema of city officials, see Becher (2014).

22. Bartram (2019a).

23. On small-time landlords and financialization, see Hulse, Reynolds, and Martin (2020).

24. Hulse, Reynolds, and Martin (2020).

25. Inspectors are not otherwise concerned with illegal activities. I observed inspectors ignoring marijuana use, public drunkenness, and possible drug deals. One inspector described an unspoken understanding between inspectors and drug dealers. "They know who [we] are by the blue shirt," he told me, "They know he's a building inspector, he's not here to interrupt trade as it were. He's just here to do his job."

26. On landlords and the threat of eviction, see Garboden and Rosen (2019).

27. Extant studies also attest to tendencies among working-class men to disdain people who complain and who "milk" the system (Lamont 2000).

28. Čapek and Gilderbloom (1992); Danielson (1976); Goetz and Sidney (1994); Krueckeberg (1999); McCabe (2016).

29. Sabbeth (2019) suggests that housing standards are not enforced because poor renters get little attention from code enforcement officials. My research builds on this and also shows that inspectors do not help tenants because they have no recourse to do so.

30. While inspector Javier mentioned the race of these tenants, the image of gentrifying tenants is racialized implicitly too, as White.

31. Lamont (2000) also found that working-class men lauded people who held down multiple jobs.

32. US Department of Housing and Urban Development (2020).
33. Whitehead (1999).

Chapter Four

1. Bartram (2019a).
2. Garb (2005); Lauster (2016); McCabe (2016); Shlay (1986).
3. Becher (2014).
4. Herbert (2018, 2021).
5. Becher (2014) also notes evaluative differences within the category of property investment, for example between long-term property owners and absentee landlords.
6. Bartram (2019a).
7. Bartram (2021).
8. Brown-Saracino (2007).
9. Keating (2008).
10. Eddie is not the only person to have such suspicions about politically motivated inspections. In 2015, for example, *The Chicago Tribune* reported that inspectors cited a food pantry for thirty-two violations to shut it down and clear some ground for the new Obama Center (Grossman 2015). Seligman (2005) also uncovers White Chicagoans' use of code enforcement against African Americans.
11. For calls for specificity when discussing neighborhood change and gentrification, see Betancur and Smith (2016); Brown-Saracino (2009, 2017); Halle and Tiso (2014); Owens (2012); Troutman (2004).
12. Smith (1979, 2005); Zukin (1987).
13. Becher (2014).
14. Seventy-three companies were found liable for at least one case of contractor fraud between 2005 and 2010 (City of Chicago 2010).
15. Goffman (1974); Small, Harding, and Lamont (2010).
16. See Small, Harding, and Lamont (2010) for an overview of social scientific accounts of the relationship between poverty and culture.
17. Wilson and Kelling (1982).
18. The frames of inspectors of color may be an example of symbolic domination, in which the dominated see the world through the lenses of the powerful (Bourdieu 2013).
19. McCall (2013). These overlapping frames—along with inspectors' stabs at justice on behalf of White homeowners—suggest that inspectors do not act on White savior paternalism, which treats people of color as helpless or in need of help from Whites—see Hughey (2012).
20. Massey and Denton (1993).
21. My fieldwork leads me to believe that all inspectors possess White blind spots. However, I also acknowledge that inspectors of color may not have felt comfortable expressing their thoughts on White privilege to me, a White researcher.

22. As Mueller (2017) argues, even when Whites are aware of structural racism, they can be "willfully colorblind" by introducing alternative factors to either ignore or mis-analyze racial dynamics. Also see Bonilla-Silva (2010).

23. On other raced-based patterns in perceptual blind spots (specifically in terms of knowledge about neighborhoods among homeseekers), see Krysan and Bader (2009).

24. Bartram (2021).

25. Harris (2013).

26. Harris (2013).

27. Bartram (2019a).

28. Neither overlapping frames nor White blind spots necessarily equate to inspectors recording more violations in communities of color.

29. Although there are actually more 311 requests in communities of color, my statistical models probably still underestimate the number of issues in these communities versus White neighborhoods. Due to fear of violence or discrimination, people of color may be less likely to call on government services—Kirk and Papachristos (2011); Sampson and Bartusch (1998). This would mean fewer 311 requests about building issues in communities of color compared to White neighborhoods, irrespective of the condition of housing. Let's say there is a rental property whose landlord refuses to deal with bedbugs and leaky windows. If that building is in a Black, Latinx, or Asian neighborhood, one tenant might call 311 to report their landlord. But if the building is in a White neighborhood, two or three tenants might call 311. As such, there would be fewer calls in communities of color than in White neighborhoods, but inspectors would still be called to properties. Even though there are more requests in an area, the number may be disproportionate to issues.

30. It is difficult to account for housing conditions in statistical analyses (and thus to determine if inspectors' decisions mirror actual conditions) because most issues are not captured in survey data.

31. Polletta (2003); Goffman (1983).

32. On material objects' capacities to afford some actions and not others, see Domínguez Rubio (2014); Gibson (1979); McDonnell (2016).

Chapter Five

1. See Adamkiewicz et al. (2011); Grineski and Hernández (2010); Hernández (2016); Swope and Hernández (2019).

2. All states, except for Arkansas, recognize the warranty of habitability, which was instituted because code enforcement had proven so ineffective as a means for redress of poor conditions in rental housing. Some states allow tenants to stop paying rent, while other states require a tenant to make rent payments to a court or pay it into an escrow account—see Summers (2020).

3. Summers (2020); Super (2011).

4. Desmond (2016).

5. Lawyers' Committee for Better Housing (2019).

6. If she fixed none, my model predicts that the rent would be $774, based on the average increase for that tract (Bartram 2019b).

7. Institute for Housing Studies (2019).

8. See Brower (2010) on the relationship between higher rent rates and habitability laws.

9. Garboden and Newman (2012).

10. At the time of writing, landlord associations in the city were organizing to ensure that this remains the case.

11. Lawyers' Committee for Better Housing (2019).

12. Woldoff, Morrison, and Glass (2016); Zuk and Chapple (2016).

13. Taylor (2019).

14. Boeing and Waddell (2016); Desmond (2016); Desmond and Wilmers (2019). Desmond and Wilmers (2019) find that landlords face financial risks when they rent in poor communities. However, while they do pay more for maintenance costs, owing to aging housing stock or overcrowding as well as frequent missed payments and higher turnover rates, losses are rare, and landlords typically realize the surplus risk charge as higher profits.

15. For a discussion of negative credentials beyond the context of property, see Pager (2007).

16. Kohler-Hausmann (2018); Pager (2007).

17. Dewar, Seymour, and Druță (2015).

18. While other digital records, such as eviction records or criminal records, tend to stem from punitive logics of surveillance, posting building violations online stems from a progressive mandate of transparency and accountability—see City of Chicago (2012). However, this does not mean that these databases live up to expectations or promises. In 2017, the *Chicago Tribune* reported that the city of Chicago paid out about $670,000 in the previous year in response to allegations that city officials violated open records law by not providing data—Pratt (2017).

19. It is possible that fixing the violations would have cost more than this price drop, but the effects my models suggest hold for all kinds of violations, not just those that are costly to repair.

20. Bartram (2019b).

21. My statistical models use the average property price in an area to control for price variations that stem from locale.

22. In Chicago, the median property value for White homeowners is $275,000, compared to $145,000 and $180,000 for Black and Latinx homeowners, respectively (Institute for Research on Race & Public Policy 2016).

23. Institute for Research on Race & Public Policy (2016); Joint Center for Housing Studies (2017).

24. Bartram (2019b).

25. On land value, see the work of rent theorists such as Harvey (2006 [1982]); Marx ([1867] 1967); Ricardo (1891); Smith ([1776] 2010); and Ward and Aalbers (2016).

26. Galanter (1974).

27. Galanter (1974).

28. Per Chicago legislation, properties owned by LLCs must be represented by a lawyer, which sets them up with better chances of avoiding fines and getting cases dismissed on their terms — Seron et al. (2001).

29. Harris (2016).

30. Kohler-Hausmann (2018).

31. Also see Feeley (1979). Building court cases are drawn out when property owners are not completing mandated repair work. This is subtly different from criminal justice settings in which delays and hassle are imposed from the top down, regardless of the behavior of the defendant.

32. Auyero (2012); Sullivan (2017).

33. Bumiller (1992); Edelman and Stryker (2005); Ewick and Silbey (1998); Galanter (1974); Rosenberg (1991); Stryker (2003, 2006, 2007).

34. Bottero explains that "people's awareness of inequality is generally framed in terms of its immediate practical relevance to their lives, as they tend to focus on those aspects of their social location which they feel can be changed, with most structures of inequality seen as beyond this scope. But while people always have partial, situated perspectives on their social world shaped by their practical engagements with it, these practical engagements vary enormously" (2019, 21). Similarly, being aware of inequality does not necessarily equate to being critical of all structures of power or systems of oppression. If a person campaigns against segregation, for example, that does not guarantee that they will support unions. Likewise, being aware of sexism in society does not equate to antiracism. Perceptions of injustice are often uneven, fragmented, and context-dependent.

35. Watkins-Hayes (2009a).

36. Stuart (2016).

37. Portes (2000); Simmel (1950). On the unintended consequences of rental regulations, see Greif (2018).

38. Young (2011).

Conclusion

1. Chicago Community Trust (2016); Istre et al. (2001); Shai (2006).

2. Becher (2014) is a notable exception.

3. Scott (1985).

4. Choi et al. (2019).

5. On informal versus formal redistribution, see Holland (2017).

6. Eubanks (2018, 7).

7. Meemoori Research AB (2018).

8. For recent overviews of rent regulations, see Rajasekaran, Treskon, and Greene (2019); Taylor (2020).

9. Garboden and Newman's (2012) study of small low-end rental housing across the US finds that preserving small rental buildings and keeping them affordable to low-income renters is only feasible with government subsidies.

10. Any increases need to be approved and are absorbed by the government unless the landlord is already charging the highest amount under the payment standard, in which case the tenant could opt to stay and make up the difference, up to 40 percent of their income.

11. Rosen (2020).

12. Hackworth (2014).

13. Blessing (2016); Fields and Uffer (2016); Robinson (2016).

14. A skeletal version of this exists in building court but is dependent on the judge appointing a receiver to cover tenants' moving expenses.

15. Summers (2020).

16. Thirty-seven of the forty biggest US cities have a program to assist low-income homeowners with home repairs, per a search of city government websites.

17. The average amount spent on repairs per property is approximately $15,355. Personal communication with staff member at Chicago Department of Planning and Development.

18. City of Chicago (2017b).

19. These conversations are part of ongoing research.

20. Garboden and Rosen (2019); Gee (2010); Monsma and Lempert (1992); Mosier and Soble (1973); Seron et al. (2001).

21. Lui (2020).

22. Smith (1979; 2005); Zukin (1987).

23. Coates (2016).

24. Mock (2021).

25. Howell and Korver-Glenn (2018); Korver-Glenn (2021).

26. The US government could also learn from the remedial efforts in South Africa, which fuse reparations with restorative justice (see Atuahene 2014).

27. Taylor (2019).

28. Marcuse and Madden (2016).

29. On community approaches to property, such as land banks and community land trusts, see Gillis (2019); Hackworth (2014); and Moore and McKee (2012).

30. Becher (2014); Herbert (2018, 2021).

31. The relationship between homeownership and safety nets may well be mutually reinforcing. Conley and Gifford's (2006) cross-national study reveals an inverse relationship between social welfare programs and homeownership.

32. Hopkins and Reyes (2021).

Appendix A

1. The demographics of my interviewees are also representative of building code inspectors nationally (International Code Council 2014).
2. Jerolmack and Khan (2014). Lamont and Swidler (2014) provide a critical overview of debates about interviews and ethnography.
3. Nagle (2013); Proudfoot and McCann (2008); Sommers (2016).
4. Eliasoph (1999); Myers (2005); Picca and Feagin (2007).
5. For more details about these methods, see Bartram (2019a).
6. Bartram (2019a).
7. I obtained rental listing data from Midwest Real Estate Data, a real estate data aggregator and distributor. The data are limited to properties listed by realtors. The benefit of this data, however, is that they list unit numbers. Other data (e.g., webscraping Craigslist listings) rarely include unit numbers (or addresses) and would not allow me to ensure that I was capturing changes in rent for the same unit in a building. For property transactions in Chicago, I used the CoreLogic database. To ensure that I compared the same unit across time, I only included condo transactions for which the data specified the unit number.
8. Bartram (2019b).
9. On fixed effects, see England et al. (1988).
10. Bartram (2019b).
11. Bartram (2019b).
12. Chicago Office of the Inspector General (2018).

Appendix C

1. This map depicts the number of STF inspections per census tract for a ten-year period as standard deviations from the mean. ArcMap calculates the mean and standard deviation. Class breaks are created with equal value ranges that are a proportion of the standard deviation—usually at intervals of 1, ½, ⅓, or ¼ standard deviations using mean values and the standard deviations from the mean.

References

Aalbers, Manuel. 2019. "Financial Geographies of Real Estate and the City: A Literature Review." *Financial Geography Working Paper Series* 21.

Adamkiewicz, Gary, Ami R. Zota, M. Patricia Fabian, Teresa Chahine, Rhona Julien, John D. Spengler, and Jonathan I. Levy. 2011. "Moving Environmental Justice Indoors: Understanding Structural Influences on Residential Exposure Patterns in Low-income Communities." *American Journal of Public Health* 101 (S1): S238–S245. https://doi.org/10.2105/AJPH.2011.300119.

Adler, Nancy E., and David H. Rehkopf. 2008. "US Disparities in Health: Descriptions, Causes, and Mechanisms." *Annual Review of Public Health* 29: 235–52.

Ahrens, Marty. 2019. "Home Structure Fires." National Fire Protection Association Research Report, October 2019. https://www.nfpa.org/-/media/Files/News-and-Research/Fire-statistics-and-reports/Building-and-life-safety/oshomes.pdf.

Atuahene, Bernadette. 2014. *We Want What's Ours: Learning from South Africa's Land Restitution Program.* Oxford: Oxford University Press.

Atuahene, Bernadette. 2018. "'Our Taxes Are Too Damn High': Institutional Racism, Property Tax Assessment, and the Fair Housing Act." *Northwestern University Law Review* 112 (6): 1501.

August, Martine, and Alan Walks. 2018. "Gentrification, Suburban Decline, and the Financialization of Multi-family Rental Housing: The Case of Toronto." *Geoforum* 89: 124–36.

Auyero, Javier. 2012. *Patients of the State: The Politics of Waiting in Argentina.* Durham, NC: Duke University Press.

Balzarini, John. 2017. "The Attitudes and Agency of Landlords in Gentrifying Neighborhoods." Paper presented at the Annual Meeting of the Urban Affairs Association, Minneapolis, MN, April 19–22, 2017.

Banks, Patricia A. 2010. *Represent: Art and Identity among the Black Upper-Middle Class.* New York: Routledge.

Bartram, Robin. 2019a. "Going Easy and Going After: Building Inspections and the Selective Allocation of Code Violations." *City & Community* 18 (2): 594–617

Bartram, Robin. 2019b. "The Cost of Code Violations: How Building Codes Shape Residential Sales Prices and Rents." *Housing Policy Debate* 29 (6): 931–46.

Bartram, Robin. 2021. "Cracks in Broken Windows: How Objects Shape Professional Evaluation." *American Journal of Sociology* 126 (4): 759–94.

Becher, Debbie. 2014. *Private Property and Public Power: Eminent Domain in Philadelphia*. Oxford: Oxford University Press.

Beck, Brenden. 2020. "Policing Gentrification: Stops and Low-Level Arrests during Demographic Change and Real Estate Reinvestment." *City & Community* 19 (1): 245–72.

Beckett, Katherine, and Steve Herbert. 2008. "Dealing with Disorder: Social Control in the Post-Industrial City." *Theoretical Criminology* 12 (1): 5–30.

Bennett, Pamela R., and Amy Lutz. 2009. "How African American Is the Net Black Advantage? Differences in College Attendance among Immigrant Blacks, Native Blacks, and Whites." *Sociology of Education* 82 (1): 70–100.

Berrey, Ellen. 2015. *The Enigma of Diversity: The Language of Race and the Limits of Racial Justice*. Chicago: University of Chicago Press.

Besbris, Max, and Jacob William Faber. 2017. "Investigating the Relationship between Real Estate Agents, Segregation, and House Prices: Steering and Upselling in New York State." *Sociological Forum* 32 (4): 850–73.

Betancur, John, and Janet Smith. 2016. *Claiming Neighborhood: New Ways of Understanding Urban Change*. Urbana, Chicago, and Springfield: University of Illinois Press.

Bey, Lee. 2019. *Southern Exposure: The Overlooked Architecture of Chicago's South Side*. Evanston, IL: Northwestern University Press.

Black, Timuel D. 2019. *Sacred Ground: The Chicago Streets of Timuel Black*. Evanston, IL: Northwestern University Press.

Blake, Mary Kate. 2018. "All Talk and No Action? Racial Differences in College Behaviors and Attendance." *Sociological Perspectives* 61 (4): 553–72.

Blessing, Anita. 2016. "Repackaging the Poor? Conceptualising Neoliberal Reforms of Social Rental Housing." *Housing Studies* 31 (2): 149–72.

Bobo, Lawrence, and Victor Thompson. 2006. "Unfair by Design: The War on Drugs, Race, and the Legitimacy of the Criminal Justice System." *Social Research: An International Quarterly* 73 (2): 445–72.

Boeing, Geoff, and Paul Waddell. 2016. "New Insights into Rental Housing Markets across the United States: Web Scraping and Analyzing Craigslist Rental Listings." *Journal of Planning Education and Research* 37: 457–76.

Bonilla-Silva, Eduardo. 2010. *Racism without Racists: Color-blind Racism and the Persistence of Racial Inequality in America*. Lanham, MD: Rowman & Littlefield.

Bottero, Wendy. 2019. *A Sense of Inequality*. Lanham, MD: Rowman & Littlefield.

Boudon, R. 1982. *The Unintended Consequences of Social Action*. London: Macmillan.

Bourdieu, Pierre. 2013. *Distinction: A Social Critique of the Judgement of Taste*. London: Routledge.

Brower, Michael A. 2010. "The Backlash of the Implied Warranty of Habitability: Theory vs. Analysis." *DePaul Law Review* 60: 849.

Brown-Saracino, Japonica. 2007. "Virtuous Marginality: Social Preservationists and the Selection of the Old-timer." *Theory and Society* 36 (5): 437–68.

Brown-Saracino, Japonica. 2009. *A Neighborhood that Never Changes: Gentrification, Social Preservation, and the Search for Authenticity*. Chicago: University of Chicago Press.

Brown-Saracino, Japonica. 2017. "Explicating Divided Approaches to Gentrification and Growing Income Inequality." *Annual Review of Sociology* 43: 515–39.

Bumiller, Kristin. 1992. *The Civil Rights Society: The Social Construction of Victims*. Baltimore, MD: Johns Hopkins University Press.

Calame, Sarah, and Aravind Boddupalli. 2020. "Fines and Forfeitures and Racial Disparities." *Tax Policy Center*, August 14. https://www.taxpolicycenter.org/taxvox/fines-and-forfeitures-and-racial-disparities.

Čapek, Stella M., and John Ingram Gilderbloom. 1992. *Community Versus Commodity: Tenants and the American City*. Albany: SUNY Press.

Center for Municipal Finance. 2021. "Cook County Scavenger Sale Evaluation." University of Chicago Center for Municipal Finance. Accessed June 28, 2021. https://harris.uchicago.edu/files/scavenger_sale.pdf.

Centro Autónomo. 2016. *Displacement in Albany Park: Housing Hardships for Low-income Tenants*. Chicago: Centro Autónomo, Spring 2016. https://ausm.community/wp-content/uploads/2018/08/GentrificationReport-2016_0.pdf.

Centro Autónomo de Albany Park. 2020. "Albany Park Neighborhood." Accessed July 16, 2020. https://centro.community/albany-park-neighborhood/

Chicago Community Trust. 2016. "In Chicago's Highest Fire Risk Neighborhoods, Protecting Homes and Families." https://www.cct.org/2016/06/in-chicagos-highest-fire-risk-neighborhoods-protecting-homes-and-families/.

Chicago Office of the Inspector General. 2018. *Department of Buildings Complaint-Based Inspections Audit*. http://chicagoinspectorgeneral.org/wp-content/uploads/2018/04/DOB-Complaint-Inspections-Audit.pdf.

Chicago Police Department. 2012. *Municipal Drug and Gang House Enforcement Program*. http://directives.chicagopolice.org/directives/data/a7a57be2-12bcfa66-cf112-bd07-b8d41410021de0c2.html.

Choi, Jung Hyun, et al. 2019. "Explaining the Black-White Homeownership Gap: A Closer Look at Disparities across Local Markets." Urban Institute Research Report. https://www.urban.org/research/publication/explaining-black-white-homeownership-gap-closer-look-disparities-across-local-markets.

Christensen, P. H., G. Warren-Myers, A. Shirazi, and X. J. Ge. 2019. "What Are Microunits and Can This New Housing Typology Help Solve the Housing Affordability Crisis? A Review of the Literature." Paper presented at the Annual American Real Estate Society Conference, Paradise Valley, AZ, April 9–13, 2019.

Christin, Angèle. 2012. "Gender and Highbrow Cultural Participation in the United States." *Poetics* 40 (5): 423–43.

Christophers, Brett. 2020. *Rentier Capitalism: Who Owns the Economy and Who Pays for It?* London: Verso.

City of Chicago. 2010. "Contractors/Companies Found Liable of Home Repair Fraud or Consumer Fraud within the Past 5 years." Chicago: Department of Business Affairs and Consumer Protection. https://www.chicago.gov/content/dam/city/depts/bacp/licenselists/contractorsfraudhistory072610.pdf.

City of Chicago. 2012. "Open Data Executive Order (No. 2012-2)." https://www.chicago.gov/city/en/narr/foia/open_data_executiveorder.html.

City of Chicago. 2017a. "Current Employee Names, Salaries, and Position Titles." Department Human Resources. Accessed April 2, 2019. https://www.cityofchicago.org/city/en/depts/dhr/dataset/current_employeenamessalariesandpositiontitles.html.

City of Chicago. 2017b. "Roof, Porch and Emergency Heating Repair Programs." City of Chicago Department of Housing. Accessed September 9, 2018. https://www.cityofchicago.org/city/en/depts/dcd/provdrs/afford_hous/svcs/receive_emergencyhousingassistance.html.

City of Chicago. 2018. "Buildings." Accessed April 4, 2018. https://www.cityofchicago.org/city/en/depts/bldgs.html.

City of Chicago. 2019a. "Chicago 311 History." Accessed July 9, 2020. https://www.chicago.gov/city/en/depts/311/supp_info/311hist.html.

City of Chicago. 2019b. "Chicago Residential Landlord and Tenant Ordinance." Accessed April 20, 2019. https://www.chicityclerk.com/legislation-records/municipal-code.

Civic Consulting Alliance Board. 2018. *Residential Property Assessment in Cook County: Summary of Analytical Findings.* https://www.ccachicago.org/wp-content/uploads/2018/02/2018-Residential-Property-Analysis-Final.pdf.

Coates, Ta-Nehisi. "The Case for Reparations." *The Atlantic*, January 27, 2016.

Collinson, David L. 1992. *Managing the Shopfloor: Subjectivity, Masculinity and Workplace Culture.* New York: Walter de Gruyter.

Community Investment Corporation. 2018. "2018 Year End Report." Accessed April 20, 2019. http://www.cicchicago.com/wp-content/uploads/2018/12/PR-YearEnd2018.pdf.

Community Investment Corporation. 2020. "Troubled Buildings Initiative." Accessed April 20, 2019. http://www.cicchicago.com/about2/troubled-buildings.

Conley, Dalton. 2010. *Being Black, Living in the Red: Race, Wealth, and Social Policy in America.* Berkeley: University of California Press.

Conley, Dalton, and Brian Gifford. 2006. "Home Ownership, Social Insurance, and the Welfare State." *Sociological Forum* 21 (1): 55.

Connell, Raewyn. 2006. "Country/City Men." In *Country Boys: Masculinity and Rural Life*, edited by Hugh Campbell, Michael Bell, and Margaret Finney, 255–66. University Park: The Pennsylvania State University Press.

Cowan, Spencer, and Michael Aumiller. 2014. *Unresolved Foreclosures: Patterns of Zombie Properties in Cook County*. Chicago: Woodstock Institute.

Danielson, Michael N. 1976. *The Politics of Exclusion*. New York: Columbia University Press.

Department of Buildings (@ChicagoDOB). 2017. "Kicking off #BuildingSafetyMonth with List of Top 10 Building Code Violations in Chicago. #keepourbuildingssafe." Twitter post, April 2, 2019. Accessed April 20, 2019. https://twitter.com/ChicagoDOB?ref_src=twsrc%5Etfw&ref_url=https%3A%2F%2Fwww.cityofchicago.org%2Fcity%2Fen%2Fdepts%2Fbldgs.html.

Desmond, Matthew. 2016. *Evicted: Poverty and Profit in the American City*. New York: Crown.

Desmond, Matthew. 2017. "How Homeownership Became the Engine of American Inequality." *The New York Times*, May 9.

Desmond, Matthew, Carl Gershenson, and Barbara Kiviat. 2015. "Forced Relocation and Residential Instability among Urban Renters." *Social Service Review* 89 (2): 227–62.

Desmond, Matthew, and Tracey Shollenberger. 2015. "Forced Displacement from Rental Housing: Prevalence and Neighborhood Consequences." *Demography* 52 (5): 1751–72.

Desmond, Matthew, and Bruce Western. 2018. "Poverty in America: New Directions and Debates." *Annual Review of Sociology* 44: 305–18.

Desmond, Matthew, and Nathan Wilmers. 2019. "Do the Poor Pay More for Housing? Exploitation, Profit, and Risk in Rental Markets." *American Journal of Sociology* 124 (4): 1090–1124.

Dewar, Margaret, Eric Seymour, and Oana Druță. 2015. "Disinvesting in the City: The Role of Tax Foreclosure in Detroit." *Urban Affairs Review* 51 (5): 587–615.

De Zwart, Frank. 2015. "Unintended but not Unanticipated Consequences." *Theory and Society* 44 (3): 283–97.

Doering, Jan. 2016. "Visibly White: How Community Policing Activists Negotiate their Whiteness." *Sociology of Race and Ethnicity* 2 (1): 106–19.

Domínguez Rubio, Fernando. 2014. "Preserving the Unpreservable: Docile and Unruly Objects at MoMA." *Theory and Society* 43 (6): 617–45.

Dwyer, Rachel E. 2018. "Credit, Debt, and Inequality." *Annual Review of Sociology* 44: 237–61.

Edelman, Lauren B., and Robin Stryker. 2005. "A Sociological Approach to Law and the Economy." In *The Handbook of Economic Sociology,* edited by N. J. Smelser and R. Swedberg, 526–51. Princeton, NJ: Princeton University Press.

Eliasoph, Nina. 1999. "'Everyday Racism' in a Culture of Political Avoidance: Civil Society, Speech, and Taboo." *Social Problems* 46 (4): 479–502.

England, Paula, et al. 1988. "Explaining Occupational Sex Segregation and Wages: Findings from a Model with Fixed Effects." *American Sociological Review* 53: 544–58.

English, Devin, Sharon F. Lambert, Michele K. Evans, and Alan B. Zonderman. 2014. "Neighborhood Racial Composition, Racial Discrimination, and Depressive Symptoms in African Americans." *American Journal of Community Psychology* 54 (3-4): 219–28.

Epp, Charles R., Steven Maynard-Moody, and Donald P. Haider-Markel. 2014. *Pulled Over: How Police Stops Define Race and Citizenship*. Chicago: University of Chicago Press.

Eubanks, Virginia. 2018. *Automating Inequality: How High Tools Profile, Police, and Punish the Poor*. New York: Picador.

Ewick, Patricia, and Susan Silbey. 1998. *The Common Place of Law: Stories from Everyday Life*. Chicago: University of Chicago Press.

Federal Bureau of Investigation. 2009. "City Inspector Charged with Bribery in Probe of Crooked Permits." US Attorney's Office Press Release, August 10, 2009. https://archives.fbi.gov/archives/chicago/press-releases/2009/cg081009.htm.

Feeley, Malcolm M. 1979. *The Process Is the Punishment: Handling Cases in a Lower Criminal Court*. New York: Russell Sage Foundation.

Fernandez, Rodrigo, and Manuel B. Aalbers. 2016. "Financialization and Housing: Between Globalization and Varieties of Capitalism." *Competition & Change* 20 (2): 71–88.

Fields, Desiree. 2018. "Constructing a New Asset Class: Property-led Financial Accumulation after the Crisis." *Economic Geography* 94 (2): 118–40.

Fields, Desiree, and Sabina Uffer. 2016. "The Financialisation of Rental Housing: A Comparative Analysis of New York City and Berlin." *Urban Studies* 53 (7): 1486–1502.

Fishman, Robert. 1987. *Bourgeois Utopias: The Rise and Fall of Suburbia*. New York: Basic Books.

Flint, Barbara J. 1977. "Zoning and Residential Segregation: A Social and Physical History, 1910–1940." PhD diss., University of Chicago.

Fording, Richard C., Joe Soss, and Sanford F. Schram. 2011. "Race and the Local Politics of Punishment in the New World of Welfare." *American Journal of Sociology* 116 (5): 1610–57.

Gabbe, Charles J. 2015. "Looking through the Lens of Size: Land Use Regulations and Micro-Apartments in San Francisco." *Cityscape* 17 (2): 223–38.

Galanter, Marc. 1974. "Why the 'Haves' Come Out Ahead: Speculations on the Limits of Legal Change." *Law & Society Review* 8: 95–160.

Gamino, John. "Vacant and Abandoned: The Housing Crisis Lives On in the Homes It has Emptied—and Banks Aren't Taking Responsibility." *South Side Weekly,* April 14, 2015. https://southsideweekly.com/vacant-and-abandoned/.

Gamson, W. A., B. Fireman, and S. Rytina. 1986. *Encounters with Unjust Authority*. Chicago: Dorsey Press.

Garb, Margaret. 2005. *City of American Dreams: A History of Home Ownership and Housing Reform in Chicago, 1871–1919*. Chicago: University of Chicago Press.

Garboden, Philip ME, and Sandra Newman. 2012. "Is Preserving Small, Low-end Rental Housing Feasible?" *Housing Policy Debate* 22 (4): 507–26.

Garboden, Philip ME, and Eva Rosen. 2018. "Talking to Landlords." *Cityscape* 20 (3): 281–91.

Garboden, Philip ME, and Eva Rosen. 2019. "Serial Filing: How Landlords Use the Threat of Eviction." *City & Community* 18 (2): 638–61.

Garrido, Marco Z. 2019. *The Patchwork City: Class, Space, and Politics in Metro Manila*. Chicago: University of Chicago Press.

Gee, Harvey. 2010. "From Hallway Corridor to Homelessness: Tenants Lack Right to Counsel in New York Housing Court." *Georgetown Journal on Poverty Law & Policy* 17 (1): 87–102.

George, Samuel, et al. 2019. *The Plunder of Black Wealth in Chicago: New Findings on the Lasting Toll of Predatory Housing Contracts*. Durham, NC: Samuel DuBois Cook Center on Social Equity at Duke University. https://socialequity.duke.edu/sites/socialequity.duke.edu/files/The%20Plunder%20of%20Black%20Wealth%20in%20Chicago.pdf.

Gibson, James. 1979. *The Ecological Approach to Visual Perception*. Boston: Houghton Mifflin.

Gillis, Catherine. 2019. "Conceptualizing 'Productive Use': Dominant Narratives and Alternative Visions of Land Use in Detroit." PhD diss., Loyola University Chicago.

Goetz, Edward G., and Mara Sidney. 1994. "Revenge of the Property Owners: Community Development and the Politics of Property." *Journal of Urban Affairs* 16 (4): 319–34.

Goffman, Erving. 1974. *Frame Analysis: An Essay on the Organization of Experience*. Cambridge, MA: Harvard University Press.

Goffman, Erving. 1983. "The Interaction Order: American Sociological Association, 1982 Presidential Address." *American Sociological Review* 48 (1): 1–17.

Goldberg, David, and Trevor Griffey. 2010. *Black Power at Work: Community Control, Affirmative Action, and the Construction Industry*. Ithaca, NY: Cornell University Press.

Gotham, Kevin Fox. 2009. "Creating Liquidity Out of Spatial Fixity: The Secondary Circuit of Capital and the Subprime Mortgage Crisis." *International Journal of Urban and Regional Research* 33 (2): 355–71.

Gradel, Thomas J., and Dick W. Simpson. 2011. *Patronage, Cronyism and Criminality in Chicago Government Agencies*. University of Illinois at Chicago, Department of Political Science. https://pols.uic.edu/wp-content/uploads/sites/273/2018/10/ac_anticorruptionreport_4bb0d.pdf.

Greif, Meredith. 2018. "Regulating Landlords: Unintended Consequences for Poor Tenants." *City & Community* 17 (3): 658–74.

Grineski, Sara E., and Alma Angelica Hernández. 2010. "Landlords, Fear, and Children's Respiratory Health: An Untold Story of Environmental Injustice in the Central City." *Local Environment* 15 (3): 199–216.

Grossman, Ron. 2015. "Volunteers Pitch In, Try to Get South Side Food Pantry Up to Code." *Chicago Tribune*, November 21, 2015.

Hackworth, Jason. 2014. "The Limits to Market-based Strategies for Addressing Land Abandonment in Shrinking American Cities." *Progress in Planning* 90: 1–37.

Hall, Matthew, Kyle Crowder, and Amy Spring. 2015. "Neighborhood Foreclosures, Racial/Ethnic Transitions, and Residential Segregation." *American Sociological Review* 80 (3): 526–49.

Halle, David. 1984. *America's Working Man: Work, Home, and Politics among Blue Collar Property Owners*. Chicago: University of Chicago Press.

Halle, David. 1993. *Inside Culture: Art and Class in the American Home*. Chicago: University of Chicago Press.

Halle, David, and Elisabeth Tiso. 2014. *New York's New Edge: Contemporary Art, the High Line, and Urban Megaprojects on the Far West Side*. Chicago: University of Chicago Press.

Harcourt, Bernard. 2001. *Illusion of Order: The False Promises of Broken Windows Policing*. Cambridge, MA.: Harvard University Press.

Harris, Alexes. 2016. *A Pound of Flesh: Monetary Sanctions as Punishment for the Poor*. New York: Russell Sage Foundation.

Harris, Dianne Suzette. 2013. *Little White Houses: How the Postwar Home Constructed Race in America*. Minneapolis: University of Minnesota Press.

Harris, Lee. 2004. "'Assessing' Discrimination: The Influence of Race in Residential Property Tax Assessments." *Journal of Land Use & Environmental Law* 20 (1): 1–60.

Harvey, David. 2006 [1982]. *The Limits to Capital*. London: Verso.

Hendon, William. 1968. "Discrimination against Negro Homeowners in Property Tax Assessment." *American Journal of Economics and Sociology* 27 (2): 125–32.

Henricks, Kasey, and Louise Seamster. 2017. "Mechanisms of the Racial Tax State." *Critical Sociology* 43 (2): 169–79.

Herbert, Claire W. 2018. "Like a Good Neighbor, Squatters Are There: Property and Neighborhood Stability in the Context of Urban Decline." *City & Community* 17 (1): 236–58.

Herbert, Claire W. 2021. *A Detroit Story: Urban Decline and the Rise of Property Informality*. Berkeley: University of California Press.

Herbert, Steve. 1996. "Morality in Law Enforcement: Chasing 'Bad Guys' with the Los Angeles Police Department." *Law & Society Review* 30 (4): 799–818.

Hernandez, Alex. 2020. "Shooting of Firefighter the Last Straw for Albany Park Neighbors: 'It Didn't Used to Be Like This.'" *Block Club Chicago*, February 4. https://blockclubchicago.org/2020/02/04/shooting-of-firefighter-the-last-straw-for-albany-park-neighbors-it-didnt-used-to-be-like-this/.

Hernández, Diana. 2016. "Understanding 'Energy Insecurity' and Why It Matters to Health." *Social Science & Medicine* 167 (October): 1–10.

Hertz, Daniel Kay. 2018. *The Battle of Lincoln Park: Urban Renewal and Gentrification in Chicago.* Cleveland, OH: Belt Publishing.

Hinz, Greg. 2018. "Blockbuster Report: How Cook County Tax System Shafts the Little Guy." *Crains Chicago Business,* February 15.

Hirsch, Arnold. 2009. *Making the Second Ghetto: Race and Housing in Chicago 1940–1960.* Chicago: University of Chicago Press.

Holland, Alisha C. 2017. *Forbearance as Redistribution: The Politics of Informal Welfare in Latin America.* New York: Cambridge University Press.

Hollander, Jocelyn A., and Rachel L. Einwohner. 2004. "Conceptualizing Resistance." *Sociological Forum* 19 (4): 533–54.

Hong, Sounman. 2016. "Representative Bureaucracy, Organizational Integrity, and Citizen Coproduction: Does an Increase in Police Ethnic Representativeness Reduce Crime?" *Journal of Policy Analysis and Management* 35 (1): 11–33.

Hopkins, Madison, and Cecilia Reyes. 2021. "The Failures before the Fires." *Chicago Tribune,* April 23.

Howell, Junia, and Elizabeth Korver-Glenn. 2018. "Neighborhoods, Race, and the 21st-Century Housing Appraisal Industry." *Sociology of Race and Ethnicity* 4 (4): 473–90.

Hughey, Matthew W. 2012. "Racializing Redemption, Reproducing Racism: The Odyssey of Magical Negroes and White Saviors." *Sociology Compass* 6 (9): 751–67.

Hulse, Kath, Margaret Reynolds, and Chris Martin. 2020. "The Everyman Archetype: Discursive Reframing of Private Landlords in the Financialization of Rental Housing." *Housing Studies* 35 (6): 981–1003.

Hunt, D. Bradford. 2009. *Blueprint for Disaster: The Unraveling of Chicago Public Housing.* Chicago: University of Chicago Press.

Hunter, Marcus Anthony, Mary Pattillo, Zandria F. Robinson, and Keeanga-Yamahtta Taylor. 2016. "Black Placemaking: Celebration, Play, and Poetry." *Theory, Culture & Society* 33 (7-8): 31–56.

Huq, Efadul, and Stacy Anne Harwood. 2019. "Making Homes Unhomely: The Politics of Displacement in a Gentrifying Neighborhood in Chicago." *City & Community* 18 (2): 710–31.

Ibata, David. 1982. "When Labor-Condo Relations Sour, Both Sides Lose." *Chicago Tribune,* July 11.

Illinois General Assembly. 2010. "Condominium Property Act." Section 765 ILCS 605/14.5 – Distressed Condominium Property. https://www.ilga.gov/legislation/ilcs/ilcs3.asp?ActID=2200&ChapterID=62

Immergluck, Dan, and Jonathan Law. 2014. "Investing in Crisis: The Methods, Strategies, and Expectations of Investors in Single-Family Foreclosed Homes in Distressed Neighborhoods." *Housing Policy Debate* 24 (3): 568–93.

Institute for Housing Studies. 2013. "Overview of the Chicago Housing Market." *Chicago 5-Year Housing Plan Data Report 2013.* Chicago: DePaul University.

https://www.housingstudies.org/media/filer_public/2013/10/01/ihs_2013_over view_of_chicago_housing_market.pdf#page=14.

Institute for Housing Studies. 2019. "2019 State of Rental Housing in Cook County." Chicago: DePaul University. https://www.housingstudies.org/releases /state-rental-2019/.

Institute for Housing Studies. 2020. "Building Community Data Capacity: Developing a Model to Preserve Affordable Housing in Uncertain Times." Chicago: DePaul University https://www.housingstudies.org/blog/building-community-data -capacity-developing-model-/.

Institute for Research on Race & Public Policy. 2016. "A Tale of Three Cities: The State of Racial Justice in Chicago Report." The State of Racial Justice in Chicago: A Tale of Three Cities. https://stateofracialjusticechicago.com/a-tale-of -three-cities/.

International Code Council. 2014. "The Future of Code Officials: Results and Recommendations from a Demographic Survey." Accessed August 12, 2020. https://cdn-web.iccsafe.org/wp-content/uploads/membership_councils/2014 -ICC-NIBS-Study-The-Future-of-Code-Officials.pdf.

Istre, Gregory R., Mary A. McCoy, Linda Osborn, Jeffrey J. Barnard, and Allen Bolton. 2001. "Deaths and Injuries from House Fires." *New England Journal of Medicine* 344 (25): 1911–16.

Jackson, Kenneth T. 1987. *Crabgrass Frontier: The Suburbanization of the United States*. New York: Oxford University Press.

Jacobs, Jane. 1961. *The Death and Life of Great American Cities*. New York: Random House.

Jerolmack, Colin, and Shamus Khan. 2014. "Talk Is Cheap: Ethnography and the Attitudinal Fallacy." *Sociological Methods and Research* 43 (2): 178–209.

Joint Center for Housing Studies. 2017. *Improving America's Housing—Demographic Change and the Remodeling Outlook*. Cambridge, MA: Harvard University. https://www.jchs.harvard.edu/sites/default/files/harvard_jchs_2017_remodeling _report.pdf.

Kahrl, Andrew. 2015. "Investing in Distress: Tax Delinquency and Predatory Tax Buying in Urban America." *Critical Sociology* 43 (2): 199–219.

Kahrl, Andrew. 2016. "The Power to Destroy: Discriminatory Property Assessments and the Struggle for Tax Justice in Mississippi." *Journal of Southern History* 88 (3): 579–616.

Katz, Michael B. 2013. *The Undeserving Poor: America's Enduring Confrontation with Poverty*. Fully updated and revised edition. New York: Oxford University Press.

Keating, Ann Durkin. 2008. *Chicago Neighborhoods and Suburbs: An Historical Guide*. Chicago: University of Chicago Press.

Kefalas, Maria. 2003. *Working-Class Heroes: Home, Community, and Nation in a Chicago Neighborhood*. Berkeley: University of California Press.

Keister, Lisa A., and Stephanie Moller. 2000. "Wealth Inequality in the United States." *Annual Review of Sociology* 26 (1): 63–81.

Kim, Mikyong Minsun, and Clifton F. Conrad. 2006. "The Impact of Historically Black Colleges and Universities on the Academic Success of African-American Students." *Research in Higher Education* 47 (4): 399–427.

Kirk, David S., and Andrew V. Papachristos. 2011. "Cultural Mechanisms and the Persistence of Neighborhood Violence." *American Journal of Sociology* 116 (4): 1190–1233.

Klein, Woody. 1964. *Let In the Sun*. New York: Macmillan.

Klepeis, Neil E., William C. Nelson, Wayne R. Ott, John P. Robinson, Andy M. Tsang, Paul Switzer, Joseph V. Behar, Stephen C. Hern, and William H. Engelmann. 2001. "The National Human Activity Pattern Survey (NHAPS): A Resource for Assessing Exposure to Environmental Pollutants." *Journal of Exposure Science & Environmental Epidemiology* 11 (3): 231–52.

Kohler-Hausmann, Issa. 2018. *Misdemeanorland: Criminal Courts and Social Control in an Age of Broken Windows Policing*. Princeton, NJ: Princeton University Press.

Korver-Glenn, Elizabeth. 2018. "Compounding Inequalities: How Racial Stereotypes and Discrimination Accumulate across the Stages of Housing Exchange." *American Sociological Review* 83 (4): 627–56.

Korver-Glenn, Elizabeth. 2021. *Race Brokers: Housing Markets and Segregation in 21st-Century Urban America*. New York: Oxford University Press.

Kurth, Ryan. 2012. "Single-Family Rental Housing—The Fastest Growing Component of the Rental Market." *Fannie Mae: Economic and Strategic Research Data Note* 2 (1). Accessed May 4, 2018. http://www.fanniemae.com/resources/file/research/datanotes/pdf/data-note-0312.pdf.

Krinsky, John, and Maud Simonet. 2017. *Who Cleans the Park? Public Work and Urban Governance in New York City*. Chicago: University of Chicago Press.

Krueckeberg, Donald A. 1999. "The Grapes of Rent: A History of Renting in a Country of Owners." *Housing Policy Debate* 10 (1): 9–30.

Krysan, Maria, and Michael D. M. Bader. 2009. "Racial Blind Spots: Black-White-Latino Differences in Community Knowledge." *Social Problems* 56 (4): 677–701.

Krysan, Maria, and Kyle Crowder. 2017. *Cycle of Segregation: Social Processes and Residential Stratification*. New York: Russell Sage Foundation.

Lamont, Michèle. 2000. *The Dignity of Working Men: Morality and the Boundaries of Race, Class, and Immigration*. New York: Russell Sage Foundation.

Lamont, Michèle, Stefan Beljean, and Matthew Clair. 2014. "What Is Missing? Cultural Processes and Causal Pathways to Inequality." *Socio-Economic Review* 12 (3): 573–608.

Lamont, Michèle, and Ann Swidler. 2014. "Methodological Pluralism and the Possibilities and Limits of Interviewing". *Qualitative Sociology* 37 (2): 153–71.

Laniyonu, Ayobami. 2018. "Coffee Shops and Street Stops: Policing Practices in Gentrifying Neighborhoods." *Urban Affairs Review* 54 (5): 898–930.

Lara-Millan, Armando. 2017 "States as a Series of People Exchanges." In *The Many Hands of the State: Theorizing Political Authority and Social Control*, edited by Kimberly J. Morgan and Ann Shola Orloff, 81–102. New York: Cambridge University Press.

Lauster, Nathanael. 2016. *The Death and Life of the Single-Family House: Lessons from Vancouver on Building a Livable City*. Philadelphia, PA: Temple University Press.

Lawyers' Committee for Better Housing. 2019. "Opening the Door on Chicago Evictions: Chicago's Ongoing Crisis." *Chicago Evictions*. https://eviction.lcbh.org/reports/chicagos-ongoing-crisis.

Leicht, Kevin T. 2008. "Broken Down by Race and Gender? Sociological Explanations of New Sources of Earnings Inequality." *Annual Review of Sociology* 34: 237–55.

Lewis, Amanda E., and John B. Diamond. 2015. *Despite the Best Intentions: How Racial Inequality Thrives in Good Schools*. New York: Oxford University Press.

Lewis, Gail. 2000. *'Race,' Gender, Social Welfare: Encounters in a Postcolonial Society*. Cambridge: Polity Press.

Lipsky, Michael. 2010. *Street-level Bureaucracy: Dilemmas of the Individual in Public Service*. New York: Russell Sage Foundation.

Lizardo, Omar. 2006. "The Puzzle of Women's 'Highbrow' Culture Consumption: Integrating Gender and Work into Bourdieu's Class Theory of Taste." *Poetics* 34 (1): 1–23.

Lui, Ann. 2020. "Toward an Office of the Public Architect." In "Expanding Modes of Practice," edited by Bryony Roberts, special issue, *Log*, No. 48.

Lutton, Linda. 2020. "Activists Want Billions in Reparations from Chase Bank for Chicago's Black Neighborhoods." WBEZ Chicago, June 15, 2020. https://www.wbez.org/stories/activists-want-billions-in-reparations-from-chase-bank-for-chicagos-black-neighborhoods/0cca1b18-c141-4630-92d0-96cdcdc77fa5.

Mallach, Alan. 2014. "Lessons from Las Vegas: Housing Markets, Neighborhoods, and Distressed Single-family Property Investors." *Housing Policy Debate* 24 (4): 769–801.

Mallach, Alan. 2018. *The Divided City: Poverty and Prosperity in Urban America*. Washington, DC: Island Press.

Marcuse, Peter, and David Madden. 2016. *In Defense of Housing: The Politics of Crisis*. London: Verso Books.

Marx, Karl. (1867) 1967. *Capital*. 3 vols. New York: International Publishers.

Massey, Douglas S., and Nancy A. Denton. 1993. *American Apartheid: Segregation and the Making of the Underclass*. Cambridge, MA: Harvard University Press.

Maynard-Moody, Steven Williams, Michael Musheno, and Michael Craig Musheno. 2003. *Cops, Teachers, Counselors: Stories from the Front Lines of Public Service*. Ann Arbor: University of Michigan Press.

McCabe, Brian J. 2016. *No Place Like Home: Wealth, Community, and the Politics of Homeownership*. New York: Oxford University Press.

McCabe, Brian J. 2018. "Costly, Regressive & Ineffective: How Sensitive Is Public Support for the Mortgage Interest Deduction in the United States?" *Housing Policy Debate* 28 (6): 963–78.

McCall, Leslie. 2013. *The Undeserving Rich: American Beliefs about Inequality, Opportunity, and Redistribution*. New York: Cambridge University Press.

McDonnell, Terence E. 2016. *Best Laid Plans: Cultural Entropy and the Unraveling of AIDS Media Campaigns*. Chicago: University of Chicago Press.

Michney, Todd M., and LaDale Winling. 2020. "New Perspectives on New Deal Housing Policy: Explicating and Mapping HOLC Loans to African Americans." *Journal of Urban History* 46 (1): 150–80.

McKenzie, Evan. 2019. "Private Covenants, Public Laws and the Financial Future of Residential Private Governments." Paper prepared for the Workshop on the Ostrom Workshop (WOW6) Conference, Indiana University, Bloomington, June 19–21, 2019. https://dlc.dlib.indiana.edu/dlc/bitstream/handle/10535/10495/mckenzie_ostrom_061019.pdf?sequence=1&isAllowed=y.

Meemoori Research AB. 2018. "Big Data for Smart Buildings 2015 to 2020." Accessed April 4, 2018. https://www.memoori.com/portfolio/big-data-smart-buildings-2015-2020/.

Mele, Christopher. 2000. *Selling the Lower East Side: Culture, Real Estate, and Resistance in New York City*. Minneapolis: University of Minnesota Press.

Merolla, David M. 2013. "The Net Black Advantage in Educational Transitions: An Education Careers Approach." *American Educational Research Journal* 50 (5): 895–924.

Merton, Robert K. 1936. "The Unanticipated Consequences of Purposive Social Action." *American Sociological Review* 1 (6): 894–904.

Metropolitan Tenants Organization. 2010. "Chicago Residential Landlord Tenant Ordinance." https://www.tenants-rights.org/residential-landlord-tenant-ordinance/.

Mijs, Jonathan J. B. 2018. "Inequality Is a Problem of Inference: How People Solve the Social Puzzle of Unequal Outcomes." *Societies* 8 (3): 64.

Minkoff, Scott L., and Jeffrey Lyons. 2019. "Living with Inequality: Neighborhood Income Diversity and Perceptions of the Income Gap." *American Politics Research* 47 (2): 329–61.

Mock, Brentin. 2021. "What It Actually Means to Pass Local 'Reparations.'" Bloomberg *CityLab Daily*, April 16. Bloomberg. https://www.bloomberg.com/news/newsletters/2021-04-16/citylab-daily-what-it-actually-means-to-pass-local-reparations.

Monsma, Karl, and Richard Lempert. 1992. "The Value of Counsel: 20 Years of Representation before a Public Housing Eviction Board." *Law & Society Review* 26 (3): 627–68.

Moore, Natalie Y. 2016. *The South Side: A Portrait of Chicago and American Segregation*. New York: Macmillan.

Moore, T., and K. McKee. 2012. "Empowering Local Communities? An International Review of Community Land Trusts." *Housing Studies* 27 (2): 280–90.

Morgan, Kimberly J., and Ann Shola Orloff, eds. 2017. *The Many Hands of the State: Theorizing Political Authority and Social Control*. New York: Cambridge University Press.

Mosier, Marilyn Miller, and Richard A. Soble. 1973. "Modern Legislation, Metropolitan Court, Miniscule Results: A Study of Detroit's Landlord-Tenant Court." *University of Michigan Journal of Law Reform* 7 (1): 8–70.

Mueller, Jennifer C. 2017. "Racial Ideology or Racial Ignorance? An Alternative Theory of Racial Cognition." *Sociological Theory* 38 (2): 142–69.

Myers, Kristen A. 2005. *Racetalk: Racism Hiding in Plain Sight*. Lanham, MD: Rowman & Littlefield.

Mykerezi, Elton, and Bradford F. Mills. 2008. "The Wage Earnings Impact of Historically Black Colleges and Universities." *Southern Economic Journal* 75 (1): 173–87.

Nagle, Robin. 2013. *Picking Up: On the Streets and Behind the Trucks with the Sanitation Workers of New York City*. New York: Macmillan.

Nellis, Ashley. "The Color of Justice: Racial and Ethnic Disparity in State Prisons." The Sentencing Project, June 14, 2016. https://www.sentencingproject.org/publications/color-of-justice-racial-and-ethnic-disparity-in-state-prisons/.

Newman, Katherine, and E. Wyly. 2006. "The Right to Stay Put, Revisited: Gentrification and Resistance to Displacement in New York City." *Urban Studies* 43 (1): 23–57.

Novak, William J. 1996. *The People's Welfare: Law and Regulation in Nineteenth-century America*. Chapel Hill: University of North Carolina Press.

Okulicz-Kozaryn, Adam. 2019. "Are We Happier among Our Own Race?" *Economics and Sociology* 12 (2): 11–35.

Oliver, Melvin, and Thomas Shapiro. 1995. *Black Wealth, White Wealth*. New York: Routledge.

Owens, Ann. 2012. "Neighborhoods on the Rise: A Typology of Neighborhoods Experiencing Socioeconomic Ascent." *City & Community* 11 (4): 345–69.

Owens, Emiel W., Andrea J. Shelton, Collette M. Bloom, and J. Kenyatta Cavil. 2012. "The Significance of HBCUs to the Production of STEM Graduates: Answering the Call." *Educational Foundations* 26: 33–47.

Pager, Devah. 2007. "The Use of Field Experiments for Studies of Employment Discrimination: Contributions, Critiques, and Directions for the Future." *The Annals of the American Academy of Political and Social Science* 609 (1): 104–33.

Parker, Jeffrey Nathaniel. 2018. "Broken Windows as Growth Machines: Who Benefits from Urban Disorder and Crime?" *City & Community* 17 (4): 945–71.

Pattillo, Mary. 2013. *Black Picket Fences: Privilege and Peril among the Black Middle Class*. University of Chicago Press.

Pattillo, Mary. 2021. "Black Advantage Vision: Flipping the Script on the Study of Racial Inequality." *Issues in Race & Society. Issues in Race & Society* 5-39.

Pattillo, Mary, Bruce Western, and David Weiman, eds. 2004. *Imprisoning America: The Social Effects of Mass Incarceration*. New York: Russell Sage Foundation.

Picca, Leslie H., and Joe R. Feagin. 2007. *Two-faced Racism: Whites in the Backstage and Frontstage*. New York: Taylor & Francis.

Pickett, Kate E., and Richard G. Wilkinson. 2008. "People Like Us: Ethnic Group Density Effects on Health." *Ethnicity & Health* 13 (4): 321–34.

Plotkin, Wendy. 2001. "'Hemmed In': The Struggle against Racial Restrictive Covenants and Deed Restrictions in Post-WWII Chicago." *Journal of the Illinois State Historical Society* 94 (1): 39–69.

Plumer, Brad, Nadja Popovich, and Brian Palmer. 2020. "How Decades of Racist Housing Policy Left Neighborhoods Sweltering." *The New York Times*, August 24.

Polletta, Francesca. 2003. "Culture Is Not Just in Your Head." In *Rethinking Social Movements: Structure, Meaning, and Emotion*, edited by Jeff Goodwin and James M. Jasper, 97–110. Lanham, MD: Rowman & Littlefield.

Portes, Alejandro. 2000. "The Hidden Abode: Sociology as Analysis of the Unexpected." *American Sociological Review* 65 (1): 1–18.

Portillo, Shannon, and Danielle S. Rudes. 2014. "Construction of Justice at the Street Level." *Annual Review of Law and Social Science* 10: 321–34.

Pratt, Gregory. 2017. "Chicago Paid $670,000 in 2016 over Lawsuits Alleging Open Records Violations." *Chicago Tribune*, January 2, 2017. https://www.chicagotribune.com/news/ct-chicago-foia-violation-lawsuits-met-20170101-story.html.

Progress Illinois. 2015. "Protesters Speak Out against the Gentrification of Chicago's Albany Park Neighborhood." *Progress Illinois*, November 9. https://www.progressillinois.com/files/quick-hits/content/2015/11/09/protesters-speak-out-against-gentrification-chicagos-albany-park/.

Prottas, Jeffrey Manditch. 1979. *People Processing: The Street-level Bureaucrat in Public Service Bureaucracies*. Lexington, MA: Lexington Books.

Proudfoot, Jesse, and Eugene J. McCann. 2008. "At Street Level: Bureaucratic Practice in the Management of Urban Neighborhood Change." *Urban Geography* 29 (4): 348–70.

Purifoye, Gwendolyn Y. 2017. "Transporting Urban Inequality Through Public Transit Designs & Systems." *City and Community* 16 (4): 364–68.

Purser, Gretchen. 2016. "The Circle of Dispossession: Evicting the Urban Poor in Baltimore." *Critical Sociology* 42 (3): 393–415.

Quadagno, Jill S. 1994. *The Color of Welfare: How Racism Undermined the War on Poverty*. New York: Oxford University Press.

Rajasekaran, Prasanna, Mark Treskon, and Solomon Greene. 2019. "Rent Control." *The Urban Institute*. https://www.urban.org/sites/default/files/publication/99646

/rent_control._what_does_the_research_tell_us_about_the_effectiveness_of_local_action_1.pdf.

Rawlings, Craig M., and Clayton Childress. 2019. "Emergent Meanings: Reconciling Dispositional and Situational Accounts of Meaning-Making from Cultural Objects." *American Journal of Sociology* 124 (6): 1763–1809.

Reosti, Anna. 2021. "The Costs of Seeking Shelter for Renters with Discrediting Background Records." *City & Community*, May. DOI:10.1177/15356841211012483.

Ricardo, David. 1891. *Principles of Political Economy and Taxation*. Edited with an introduction by Edward Carter Kersey Gonner. London: G. Bell and Sons.

Robinson, John N., III. 2016. "Race, Poverty, and Markets: Urban Inequality after the Neoliberal Turn." *Sociology Compass* 10 (12): 1090–1101.

Rodriquez, Jason. 2006. "Color-Blind Ideology and the Cultural Appropriation of Hip-Hop." *Journal of Contemporary Ethnography* 35 (6): 645–68.

Rolnik, Raquel. 2013. "Late Neoliberalism: The Financialization of Homeownership and Housing Rights." *International Journal of Urban and Regional Research* 37 (3): 1058–66.

Rosen, Eva. 2020. *The Voucher Promise: "Section 8" and the Fate of an American Neighborhood*. Princeton, NJ: Princeton University Press.

Rosenberg, Gerald N. 1991. *The Hollow Hope: Can Courts Bring About Social Change?* Chicago and London: University of Chicago Press.

Rotella, Carlo. 2019. *The World Is Always Coming to an End: Pulling Together and Apart in a Chicago Neighborhood*. Chicago Visions and Revisions. Chicago: University of Chicago Press.

Rothstein, Richard. 2017. *The Color of Law: A Forgotten History of How our Government Segregated America*. New York: Liveright Publishing.

Rugh, Jacob S., Len Albright, and Douglas S. Massey. 2015. "Race, Space, and Cumulative Disadvantage: A Case Study of the Subprime Lending Collapse." *Social Problems* 62 (2): 186–218.

Rugh, Jacob S., and Douglas S. Massey. 2010. "Racial Segregation and the American Foreclosure Crisis." *American Sociological Review* 75 (5): 629–51.

Runciman, Garry. 1966. *Relative Deprivation and Social Justice: A Study of Attitudes to Social Inequality in Twentieth-century England*. Berkeley: University of California Press.

Sabbeth, Kathryn A. 2019. "(Under) Enforcement of Poor Tenants' Rights." *Georgetown Journal on Poverty Law & Policy* 27 (1): 97–146.

Sampson, Robert J. 2012. *Great American City: Chicago and the Enduring Neighborhood Effect*. Chicago: University of Chicago Press.

Sampson, Robert J., and Dawn Jeglum Bartusch. 1998. "Legal Cynicism and (Subcultural?) Tolerance of Deviance: The Neighborhood Context of Racial Difference." *Law & Society Review* 32 (4): 777–804.

Sampson, Robert J., and Stephen W. Raudenbush. 2004. "Seeing Disorder: Neighborhood Stigma and the Social Construction of 'Broken Windows.'" *Social Psychology Quarterly* 67 (4): 319–42.

Sampson, Robert J., and Alix Winter. 2016. "The Racial Ecology of Lead Poisoning: Toxic Inequality in Chicago Neighborhoods, 1995–2013." *DuBois Review: Social Science Research on Race* 13 (2): 261–83.

Sanchez, Melissa, and Sandhya Kambhampati. 2018. "How Chicago Ticket Debt Sends Black Motorists into Bankruptcy." *Propublica Illinois,* February 27. https://features.propublica.org/driven-into-debt/chicago-ticket-debt-bankruptcy/.

Sassen, Saskia. 2014. *Expulsions: Brutality and Complexity in the Global Economy.* Cambridge, MA: Harvard University Press.

Satter, Beryl. 2009. *Family Properties: Race, Real Estate, and the Exploitation of Black Urban America.* New York: Metropolitan Books.

Schram, Sanford F. 2000. *After Welfare: The Culture of Postindustrial Social Policy.* New York: New York University Press.

Scott, James. 1985. *Weapons of the Weak: Everyday Forms of Peasant Resistance.* New Haven, CT: Yale University Press.

Seim, Josh. 2020. *Bandage, Sort, and Hustle: Ambulance Crews on the Front Lines of Urban Suffering.* Berkeley: University of California Press.

Seligman, Amanda. 2005. *Block by Block: Neighborhoods and Public Policy on Chicago's West Side.* Chicago: University of Chicago Press.

Semuels, Alana. 2019. "When Wall Street Is Your Landlord." *The Atlantic*, February 13. https://www.theatlantic.com/technology/archive/2019/02/single-family-landlords-wall-street/582394/.

Seron, Carroll, Martin Frankel, Gregg Van Ryzin, and Jean Kovath. 2001. "The Impact of Legal Counsel on Outcomes for Poor Tenants in New York City's Housing Court: Results of a Randomized Experiment." *Law & Society Review* 35 (2): 419–34.

Shai, Donna. 2006. "Income, Housing, and Fire Injuries: A Census Tract Analysis." *Public Health Reports* 121 (2): 149–54.

Sharkey, Patrick. 2013. *Stuck in Place: Urban Neighborhoods and the End of Progress toward Racial Equality.* Chicago: University of Chicago Press.

Shedd, Carla. 2015. *Unequal City: Race, Schools, and Perceptions of Injustice.* New York: Russell Sage Foundation.

Shiffer-Sebba, Doron. 2020. "Understanding the Divergent Logics of Landlords: Circumstantial versus Deliberate Pathways." *City & Community* 15 (2): 137–62. http://doi.org/ 10.1111/cico.12490.

Shively, JoEllen. 1992. "Cowboys and Indians: Perceptions of Western Films among American Indians and Anglos." *American Sociological Review* 57 (6): 725–34.

Shlay, Anne B. 1986. "Taking Apart the American Dream: The Influence of Income and Family Composition on Residential Evaluation." *Urban Studies* 23 (4): 253–70.

Simmel, Georg. 1950. "The Stranger." In *The Sociology of Georg Simmel*, translated by Kurt Wolff, 402–8. New York: Free Press.

Singla, Akheil, Charlotte Kirschner, and Samuel B. Stone. 2020. "Race, Representation, and Revenue: Reliance on Fines and Forfeitures in City Governments." *Urban Affairs Review* 56 (4): 1132–67.

Skogan, Wesley G. 2006. *Police and Community in Chicago: A Tale of Three Cities*. New York: Oxford University Press.

Small, Mario L., David J. Harding, and Michèle Lamont. 2010. "Reconsidering Culture and Poverty." *The ANNALS of the American Academy of Political and Social Science* 629 (1): 6–27.

Smith, Adam. (1776) 2010. *The Wealth of Nations: An Inquiry into the Nature and Causes of the Wealth of Nations*. Petersfield, UK: Harriman House Limited.

Smith, Neil. 1979. "Toward a Theory of Gentrification: A Back to the City Movement by Capital, not People." *Journal of the American Planning Association* 45 (4): 538–48.

Smith, Neil. 1987. "Gentrification and the Rent Gap." *Annals of the Association of American Geographers* 77 (3): 462–65.

Smith, Neil. 1991. *Uneven Development: Nature, Capital and the Production of Space*. Oxford: Blackwell.

Smith, Neil. 2005. *The New Urban Frontier: Gentrification and the Revanchist City*. New York: Routledge.

Sneath, Sara. 2020. "'It's Not Fair': Workers in a Poor Mississippi County Pay More Tax than Trump." *The Guardian*, October 18. https://www.theguardian.com/us-news/2020/oct/18/tax-audit-earned-income-tax-credit-mississippi

Solnit, Rebecca. 2021. "Women Are Harmed Every Day by Invisible Men." *The Guardian*, March 19. https://www.theguardian.com/commentisfree/2021/mar/19/women-harmed-every-day-invisible-men.

Sommers, Rory. 2016. "Governing Incivility: An Ethnographic Account of Municipal Law Enforcement, Urban Renewal and Neighbourhood Conflict in the City of Hamilton." PhD diss., University of Guelph.

Soss, Joe, Richard Fording, and Sanford Schram. 2011. *Disciplining the Poor: Neoliberal Paternalism and the Persistent Power of Race*. Chicago: University of Chicago Press.

Soss, Joe, Sanford Schram, Richard Fording, and Linda Houser. 2009. "Deciding to Discipline: Race, Choice, and Punishment at the Frontlines of Welfare Reform." *American Sociological Review* 74 (3): 398–422.

Stegman, Michael. 1967. "Slumlords and Public Policy." *Journal of the American Institute of Planners* 33 (5): 419–24.

Sternlieb, George. 1966. *The Tenement Landlord*. New Brunswick, NJ: Urban Studies Center, Rutgers University.

Stryker, Robin. 2003. "Mind the Gap: Law, Institutional Analysis and Socioeconomics." *Socio-Economics Review* 1: 335–67.

Stryker, Robin. 2006. "Sociology of Law." In *21st Century Sociology: A Reference Handbook*, edited by Clifton D. Bryant and Dennis L. Peck, 339–52 and 662–66. Thousand Oaks, CA: Sage Publications.

Stryker, Robin. 2007. "Half Empty, Half Full, or Neither: Law, Inequality, and Social Change in Capitalist Democracies." *Annual Review of Law and Social Science* 3: 69–97.

Stuart, Forrest. 2016. *Down, Out, and Under Arrest: Policing and Everyday Life in Skid Row*. Chicago: University of Chicago Press.

Sullivan, Esther. 2017. "Displaced in Place: Manufactured Housing, Mass Eviction, and the Paradox of State Intervention." *American Sociological Review* 82 (2): 243–69.

Summers, Brandi Thompson. 2019. *Black in Place: The Spatial Aesthetics of Race in a Post-Chocolate City*. Chapel Hill: University of North Carolina Press.

Summers, Nicole. 2020. "The Limits of Good Law." *The University of Chicago Law Review* 87 (1): 145–222.

Sun, Ivan Y., and Brian K. Payne. 2004. "Racial Differences in Resolving Conflicts: A Comparison between Black and White Police Officers." *Crime & Delinquency* 50 (4): 516–41.

Super, David A. 2011. "The Rise and Fall of the Implied Warranty of Habitability." *California Law Review* 99 (2): 389–463.

Swope, Carolyn B., and Diana Hernández. 2019. "Housing as a Determinant of Health Equity: A Conceptual Model." *Social Science & Medicine* 243: 112571.

Tapp, Renee, and Kelly Kay. 2019. "Fiscal Geographies: 'Placing' Taxation in Urban Geography." *Urban Geography* 40 (4): 573–81.

Taylor, Keeanga-Yamahtta. 2019. *Race for Profit: How Banks and the Real Estate Industry Undermined Black Homeownership*. Chapel Hill: University of North Carolina Press.

Taylor, Yesim Sayin. 2020. "Rent Control Literature Review." *DC Policy Center*. Accessed April 20, 2021. https://www.dcpolicycenter.org/publications/rent-control-literature-review/.

Teresa, Benjamin F. 2016. "Managing Fictitious Capital: The Legal Geography of Investment and Political Struggle in Rental Housing in New York City." *Environment and Planning A* 48 (3): 465–84.

Theodore, Nik, and Nina Martin. 2007. "Migrant Civil Society: New Voices in the Struggle over Community Development." *Journal of Urban Affairs* 29 (3): 269–87.

Thomas, Melvin E., Richard Moye, Loren Henderson, and Hayward Derrick Horton. 2017. "Separate and Unequal: The Impact of Socioeconomic Status, Segregation, and the Great Recession on Racial Disparities in Housing Values." *Sociology of Race and Ethnicity* 4 (2): 229–44.

Tilly, Charles. 1996. "Invisible Elbow." *Sociological Forum* 11 (4): 589–601.

Tomaskovic-Devey, Donald, and Dustin Avent-Holt. 2019. *Relational Inequalities: An Organizational Approach*. New York: Oxford University Press.

Tonkiss, Fran. 2005. *Space, the City and Social Theory: Social Relations and Urban Forms*. Cambridge: Polity Press.

Travis, Adam. 2019. "The Organization of Neglect: Limited Liability Companies and Housing Disinvestment." *American Sociological Review* 84 (1): 142–70.

Troutman, Parke. 2004 "A Growth Machine's Plan B: Legitimating Development when the Value-free Growth Ideology Is under Fire." *Journal of Urban Affairs*, 26: 611–22.

US Census Bureau. 2018. Quick Facts: Chicago City, Illinois. https://www.census.gov/quickfacts/fact/table/chicagocityillinois/DIS010218.

US Department of Housing and Urban Development. 2020. "Housing Choice Voucher Fact Sheet." https://www.hud.gov/topics/housing_choice_voucher_program_section_8.

United States Department of Justice. 2012. Press Release, July 24, 2012. "Seven Defendants, Including Three Loan Originators, Indicted in Alleged $8.5 Million Mortgage Fraud Scheme." https://archives.fbi.gov/archives/chicago/press-releases/2012/seven-defendants-including-three-loan-originatorsindicted-in-alleged-8.5-million-mortgage-fraud-scheme.

Vale, Lawrence J. 2013. *Purging the Poorest: Public Housing and the Design Politics of Twice-Cleared Communities*. Chicago: University of Chicago Press.

Valverde, Mariana. 2011. "Seeing Like a City: The Dialectic of Modern and Premodern Knowledges in Urban Governance." *Law & Society Review* 45 (2): 277–313.

Valverde, Mariana. 2012. *Everyday Law on the Street: City Governance in an Age of Diversity*. Chicago: University of Chicago Press.

Van Cleve, Nicole Gonzalez. 2016. *Crook County: Racism and Injustice in America's Largest Criminal Court*. Palo Alto, CA: Stanford University Press.

Van Maanen, John. 1978. "The Asshole." In *Policing: A View from the Street*, edited by P. K. Manning and J. Van Maanen, 221–38. Santa Monica, CA: Goodyear Publishing Company.

Vargas, Robert. 2016. *Wounded City: Violent Turf Wars in a Chicago Barrio*. New York: Oxford University Press.

Vogt Yuan, Anastasia S. 2007. "Racial Composition of Neighborhood and Emotional Well-being." *Sociological Spectrum* 28 (1): 105–29.

The Voorhees Center. 2001. *Gentrification in West Town: Contested Ground*. Accessed on July 1, 2021. https://voorheescenter.red.uic.edu/wp-content/uploads/sites/122/2017/10/Gentrification-in-West-Town-Contested-Ground.pdf.

Wacquant, Loïc. 2009. *Punishing the Poor: The Neoliberal Government of Social Insecurity*. Durham, NC: Duke University Press.

Ward, Callum, and Manuel B. Aalbers. 2016. "Virtual Special Issue Editorial Essay: 'The Shitty Rent Business': What's the Point of Land Rent Theory?" *Urban Studies* 53 (9): 1760–83.

Watkins-Hayes, Celeste. 2009a. *The New Welfare Bureaucrats: Entanglements of Race, Class, and Policy Reform*. Chicago: University of Chicago Press.

Watkins-Hayes, Celeste. 2009b. "Race-ing the Bootstrap Climb: Black and Latino Bureaucrats in Post-reform Welfare Offices." *Social Problems* 56 (2): 285–310.

Watkins-Hayes, Celeste. 2011. "Race, Respect, and Red Tape: Inside the Black Box of Racially Representative Bureaucracies." *Journal of Public Administration Research and Theory* 21 (Supplement 2): 233–51.

Weber, Rachel. 2015. *From Boom to Bubble: How Finance Built the New Chicago*. Chicago: University of Chicago Press.

Wellman, David T. 1993. *Portraits of White Racism*. New York: Cambridge University Press.
White, Kellee, and Jourdyn A. Lawrence. 2019. "Racial/Ethnic Residential Segregation and Mental Health Outcomes." In *Racism and Psychiatry: Contemporary Issues and Interventions*, edited by Morgan M. Medlock, Derri Shtasel, Nhi-Ha T. Trinh, and David R. Williams, 37–53. Cham, Switzerland: Humana Press.
Whitehead, Colson. 1999. *The Intuitionist*. New York: Anchor Books.
Wilkins, Vicky M., and Brian N. Williams. 2008. "Black or Blue: Racial Profiling and Representative Bureaucracy." *Public Administration Review* 68 (4): 654–64.
Wilson, James Q., and George L. Kelling. 1982. "Broken Windows." *Atlantic Monthly* 249 (3): 29–38.
Wilson, Valerie Rawlston. 2007. "The Effect of Attending an HBCU on Persistence and Graduation Outcomes of African-American College Students." *The Review of Black Political Economy* 34 (1): 11–52.
Wilson, William J. 1996. *When Work Disappears: The World of the New Urban Poor*. New York: Vintage Books.
Woldoff, Rachael, Lisa Morrison, and Michael Glass. 2016. *Priced Out: Stuyvesant Town and the Loss of Middle-Class Neighborhoods*. New York: New York University Press.
Woodstock Institute. 2014. "Unresolved Foreclosures: Patterns of Zombie Properties in Cook County." Accessed August 2, 2018. https://www.woodstockinst.org/research/reports/unresolved-foreclosures-patterns-zombie-properties-cook-county/.
Young, Iris Marion. 2011. *Justice and the Politics of Difference*. Princeton, NJ: Princeton University Press.
Zacka, Bernardo. 2017. *When the State Meets the Street*. Cambridge, MA: Harvard University Press.
Zuk, Miriam, and Karen Chapple. 2016. "Housing Production, Filtering and Displacement: Untangling the Relationships." UC Berkeley Research Brief, Institute for Intergovernmental Studies. https://escholarship.org/content/qt7bx938fx/qt7bx938fx.pdf.
Zukin, Sharon. 1987. "Gentrification: Culture and Capital in the Urban Core." *Annual Review of Sociology* 13 (1): 129–47.
Zukin, Sharon. 1998. "Urban Lifestyles: Diversity and Standardisation in Spaces of Consumption." *Urban Studies* 35 (5-6): 825–39.

Index

affordable housing, 40, 45–46, 72, 90, 94, 108, 153–54, 189–90n69; stacked deck, 133

African Americans, 77, 79, 81, 95–96, 123, 186n21, 186n31; campaigns of terror against, 28; contract sales, 28–29; decimation of Black neighborhoods, 120; discrimination against, 28; predatory inclusion, 29–30; stereotypes about, 121

agency: and structure, 161

American Community Survey, 171, 174–75, 187n2

American Dream, 100, 161, 192n12

apartment buildings, viii, 37–42, 44, 48, 61, 75, 77, 79, 85, 132, 138; housing vouchers, 36; rehabbed, 107

Becher, Debbie, 102–3

"blighted" neighborhoods, 42

Bottero, Wendy, 11, 186n33, 200n34

broken windows, 6, 15, 77, 83; as criminological theory, 122, 138

Brown-Saracino, Japonica, 11, 108

building code enforcement, 14, 19, 48, 72, 74, 144–45, 149, 158, 160–61, 173; property, effects on, 174–75

building code regulations, 62, 74, 130, 131, 144, 149, 152–53, 173, 193n35; enforced unevenly, 14–15; enforcement of, as punitive, 134; growth of, 61; as malleable, 167

building code violations, 2, 20, 27, 44, 46, 56, 58, 61–62, 66, 76–77, 80, 83, 90, 103, 127–28, 135, 138, 171, 178; administrative hearings, 69, 72; building court, 19, 69–72; building owner, onus on, 68; in communities of color, 49, 52, 125–26; crime fighting, 91; fixing of, and property values, 139; listed online, 67; on North Side, 49, 52–53; on Northwest Side, 53–54; property prices, effect on, 136–37; rental hikes, 131–34; rental prices, 131–34; requests, 53; on South and West Sides, 49, 52; on Southwest Side, 53; as subjective, 167; 311 requests, 53, 177

building court, 1–2, 21, 33, 39, 67, 80, 87, 94–95, 108, 114, 126, 129, 132–33, 135, 144, 154, 169, 173–77, 195n57, 200n31, 201n14; duration of cases, 128; fines, 140; justice blockers, 156; one-shotters, 140; procedural hassles, 140–42; repeat players, 140; stabs of justice, stymied in, 139; waiting, 142–43

building inspectors, 3, 9, 12–14, 18, 33, 43, 61–63, 75, 81, 128, 131, 136, 160, 165, 171, 173–74, 187n49, 192–93n19, 194n44; assessments of buildings, 98; bad landlords, 78–80, 96; benevolence of, 16–17; big buildings, 84–86; bribery and corruption, 65–66; building code violations, 49, 52–54, 56, 67–68, 91; building profit margins, assessment of, 83; building size, 83; clout of, 55–56, 60; of color, 197n18, 197n21; communities of color, benevolence toward, 60, 100; communities of color, as precarious, 124; communities of color, sympathy for, 172; compassion of, 126; condo fraud, 115–16; contexts, importance of, 112; cutting corners, disdain of,

building inspectors (*cont.*)
66, 82; decimation of Black neighborhoods, 120; decision-making of, 78, 89; defensible disrepair, 104; demographics of, 192n8; deserving homeowners, 111; deserving and undeserving tenants, distinctions between, 78, 93; dilapidation, 119; discretion, exercise of, 53–54, 56, 64, 71, 74, 77, 95, 127; disparity, motivated by, 16; disproportionate profits, assessment of, 81; ethnicity of, 15; eviction court, 155; as expert witnesses, 70; exploitation, disdain of, 80; fly-by-night contractors, 116; foreclosure, 2; foreclosures, and vacant properties, 117; and framing, 119, 121–22, 197n18–19; "good" gentrification, 107–8; greed, disdain for, 100, 112; hipsters, attitude toward, 95; homeowners, protection of, 101; and homeownership, 102–3, 114–15; housing voucher recipients, 96; illegal activities, less concerned with, 196n25; inequality, attitude toward, 96; informal training, 58; injustice, motivation by, 15; inspections, as objective and impartial, 166–67; judgment calls, 64; landlord negligence, 132; landlords, disdain toward, 94; landlords, penalizing of, 133; leniency of, 1, 15–16, 20, 54, 56, 71–72, 74, 98, 100, 120, 126, 134, 141, 143; low-income property owners, compassion toward, 103, 105; malign neglect, 89; as mostly White working-class men, 15, 54–55; motivations of, 48, 53–54, 60, 74, 94; multi-unit buildings, 83–84; and negligence, 78–79; neighborhood old-timers, protection of, 106–10, 112, 126; "no action," 67; overlooking minor issues, 71, 88, 138, 143, 150; owner-occupied buildings, stabs of justice on behalf of, 102; patterns of, 48, 52; political correctness, striving for, 57; precarity, assumptions about, 72, 100, 103, 112; precarity, and profit, 118, 126; professional landlords, 83–85; professional landlords, penalizing of, 60; property owners, 70–72, 102; as proud of their jobs, 55; punishment by, 16–17, 20, 54, 56, 71–72, 74, 98; racism, concern about, 57; receivership, contempt for, 113–14; relative assessments, 98; rental buildings, 89; rental buildings, negligence of, 78; reparative and retributive justice, attempts at, 15–16; residential violations, 191n4; ride-alongs with, 168–71, 177–78; single-family homes, protection of, 99–100, 102; small rental buildings, concern for, 88; small rental buildings, as precarious, 84; small-time landlords, 86–87, 89–90, 93; small-time landlords, leniency toward, 52, 60, 87; social location, 20, 54, 60; stabs at justice by, 17, 20–21, 53, 74, 80, 88–89, 93–94, 97–98, 100, 101, 107, 114, 116, 125, 127, 130, 135, 137, 139, 141, 144, 150–52; stacked deck, 47, 53–54, 58–60, 77, 94, 97, 112, 120, 133, 145–47; stacked deck, perceptions of, 16, 52, 93, 126–27; stacked deck, as tool, 52; starting salary of, 192n13; Strategic Task Force (STF) inspections, 90–93; struggling homeowners, leniency toward, 60, 103–5; and tenants, 97, 134, 177; tenants, lack of empathy toward, 52, 59, 88, 93–95; 311 requests, 49, 52–54, 64–66, 76, 88, 93–96, 99–100, 106, 169, 176–77, 195n59; urban profiteers, penalizing of, 60; vacant properties, and insurance companies, 117; virtuous marginality, 108–9; wealth, penalizing of, 124; White blind spots of, 17, 52, 59, 122, 197n21; Whiteness, lack of attention to, 59; White privilege, lack of attention to, 59; White savior paternalism, 197n19; working-class sensibilities of, 59. *See also* frontline workers

Buildings Department, 2, 18, 27, 48, 60–63, 73, 75–76, 78–79, 92, 165–66, 174, 176–78, 193n31, 194n40, 194n43, 194n45; bribery and corruption, 65; rotation system, 58; 311 requests, prioritizing of, 194n45; 311 system, 64–65

building violations, 18–19, 28, 34, 37–38, 56, 67, 69, 77, 127, 131, 134, 139, 170–73, 191n4, 194n45, 195n57, 199n18; in communities of color, 125–26; geography of, 49; on North Side, 52–53; owner-occupied buildings, 100; property liens, 32; property prices, 136–37, 174–75; racial disparities, in housing conditions, 125; rental prices, 132, 174; rental properties, 84; on Southwest Side, 53; stabs of justice, 53, 79; on West Side, 52

INDEX

bungalows, 36, 60, 79, 81, 109–10, 128–29, 148, 168–69; Bungalow Belt, 59

changing neighborhoods, 105–6, 111, 126, 128–29; precarity of, 110. *See also* neighborhood change
Chase Bank, 32
Chicago (Illinois), 1–2, 14, 20, 58, 68, 70, 97, 137–39, 146–47, 150, 153, 155–56, 160, 175; administrative hearings, for building code violations, 69; affordable housing, lack of, 72; Asian population, 81, 125; Black Belt, 28; Black population, 23, 28–30, 32, 34–35, 46, 58–60, 81, 112, 120, 125, 151, 157, 198n29; building code enforcement, 72; building code of, as unique, 193n37; building regulations, 61; Bungalow Belt, 59; as city of neighborhoods, 148; corruption, 65; Data Portal, 67, 171, 174; economic segregation, 103; Englewood, 169; evictions, 45; "family building," 27; fines and forfeitures, revenue from, 193n32; fire safety, concerns over, 61; geography of, 23; geography of, and stacked deck, 36; housing conditions, 22–23; housing landscape, as physical manifestation of stacked deck, 16; housing vouchers, 133, 154; intergenerational wealth and racial disparity, 34; landlordism, 40; Latinx population, 23, 30, 32, 35, 45, 59–60, 79, 81, 94–95, 103, 106, 109–11, 120, 125, 128–29, 157, 198n29; median household income, 23; median property values, 199n22; median rent in, 40; Near South Side, 120; Near West Side, 64; North Side, 19, 23–24, 32–36, 39, 46, 49, 52–53, 59, 75, 78–79, 83, 90, 107, 114, 119, 135, 143–44, 170; Northwest Side, 18–19, 23, 39, 42–46, 53, 59–60, 81, 94–95, 101; open records law, violation of, 198n18; porches, construction of, 61; public transit, 5; racial covenants, 28; racial demographics, changes in, 28; racial and economic inequality, 72, 120; racial segregation, 72; racism in, 192–93n19; and rent, 131–32; rental buildings, 84–85, 88; rental stock, 196n17; rent hikes, 133; residential shift, 28; segregation in, 23–25, 188–89n37; "side" terminology of, 23–24; 606 Trail, 106; South Loomis Boulevard, 27–33; South Shore Drive, 37–42, 44; South Side, 18–19, 23, 28, 32, 36, 39, 41, 43, 46, 49, 52–53, 59–60, 85, 119, 123, 169–70; South Side Irish, 80; Southwest Side, 19, 23, 27–30, 52–53, 79, 109, 168–69, 187n12, 188n25; stacked deck, 46–47, 72; 311 service requests, 49, 63, 176; traffic tickets, 73; Victoria Street, 33–35; Wellington Street, 44–46; West Side, 18–19, 23, 28, 32, 36, 39, 43, 46, 49, 52, 59–60, 88, 104, 110–11, 123, 128, 170; White hipsters, 95; White population, 23, 28, 32, 34–35, 40, 75, 79, 81, 83, 95, 101, 106, 110, 112, 121, 128–29, 135, 151, 157, 172, 198n29; White rehabbers, and conservation movement, 34–35; White wealth, 34; zoning decisions, 34
Chicago Community Development Financial Institutions, 189–90n69
Chicago Housing Authority (CHA), 66, 194n40
colorblindness, 124, 198n22
communities of color, 5–6, 11, 15–17, 21, 28–29, 34–35, 58, 60, 95, 119, 126, 188–89n37; building code violations, 171–72; conventional loans, denial of, 32; disinvestment, 150; disparities, 151; over-assessing of homes, 32; and poverty, 124; as precarious, 124; stacked deck, 120; 311 requests, 198n29; wealth extraction, 100
Community Initiatives, Inc. (CII), 189–90n69
condos, 38–40, 44, 64, 128–29; associations, 112; and fraud, 30
contract sales, 28–29
Cook County Tax Assessor, 174; property assessments, as subjective and discretionary, 73
court inspectors, 128; procedural hassles, 140–42; punitive stabs at justice, 129–30, 140, 143–44; stacked deck, 129–30
criminal justice system, 131, 149–50
cronyism, 114
"culture of poverty," 121; structural factors, 119

deconversion, 39
defensible disrepair, 104
Denton, Nancy, 123
Department of the Treasury, 189–90n69
Desmond, Matthew, 199n14

developers, 6–7, 15, 37, 40, 42, 45, 67, 83, 88, 106, 110, 128–29, 136–37, 149, 151, 157
dilapidation, 19, 31–33, 47, 77, 82, 119, 129–30, 145, 150, 157, 173; in Black neighborhoods, 121
disinvestment, 36, 53, 77, 79, 111–12, 123; in Black neighborhoods, 120; White blind spots, 124
disparity, 10, 16, 18
displacement, 3, 22, 42, 94–95, 130, 155, 157; Strategic Task Force-induced evictions, 44; targeted inspections, and drug and gang activity, 43

empowerment, 9
Eubanks, Virginia, 152
evictions, 22, 81, 94, 134, 154–55, 195n57; displacement, as form of, 44, 95; and poverty, 45; and racism, 131

Federal Bureau of Investigation (FBI), 39, 65, 193n20
Federal Home Loan Bank of Chicago, 189–90n69
Federal Housing Administration (FHA), 5–6, 28
Flint (Michigan), 66
flipping, 39, 41, 82, 140
foreclosures, 22, 32, 52, 79, 117, 134, 188n22, 188n28; in Black and Latinx communities, 188n27; building inspectors, 2, 17; housing conditions, 31; national banks, 118–19; property owners, 156–57; on South Side, 85; on Southwest Side, 30, 53, 188n25; zombie properties, 31, 188n30
frontline workers, 16, 58, 73, 103, 149, 153, 159–60, 192n12; automated decisions, 152; building code enforcement, 72; and discretion, 13, 56, 71, 152; and inequality, 14–15; justice blockers, 145–46, 151–52; racial bias, 170–71; residents of color, as disproportionately sanctioned, 15; stabs at justice, 97–98, 144, 146; stacked deck, reliance on, 14; tools of, 146, 150–51; urban governance, 13, 17

Galanter, Marc, 140
gentrification, 11, 94–95, 107, 111, 155, 196n30; as term, 42

Goffman, Erving, 4
Great Chicago Fire, 61

Hackworth, Jason, 154
Harris, Dianne, 35, 124
healthcare, 149–50, 158–59
Herbert, Claire, 102
heterogeneity, 23
historically black colleges and universities (HBCUs), 7
Home Affordable Refinance Program (HARP), 104
home-building industry, 35
homeownership, 126, 152, 159; American Dream, crowning point of, 100; Black homebuyers, 187n12; defensible disrepair, 103–5; dilapidation, 157; displacement, 157; home repairs, 156; owner-occupied buildings, 102; as precarious, 115, 117; property markers, 137, 143; protection of, 101–2; stacked deck, 124, 155; struggling, 102–5; valorization of, 100; and wealth, 158; White homeowners, built environment of, 35
house fires, 187n45
housing conditions, 14, 60, 100, 122–23, 148, 160–61; and foreclosure, 31; racial disparities in, 125, 130, 157, 172; racial inequity, 158; stacked deck, 47, 49
housing crash (2008), 21, 115–16, 119, 150
housing inequality, 17, 21, 47, 127; and precarity, 45–46
housing market, 131, 146–47; discrimination, 60; home warranty program, 157; racial and economic inequality, 139; real estate investment trusts, 158, 188n19; securitization, 158, 188n19; stabs at justice, obstruction of, 145
housing vouchers, 41–42, 97–98, 133, 154
Hunter, Marcus Anthony, 186n21
Hurricane Katrina, 148

illegal conversions, 70, 93, 105, 109–10, 129, 135
Illinois Circuit Court, 174
Illinois Distressed Condominium Property Act, 189n67
Illinois Urban Community Conservation Act (1953), 194n42
inequality, 3–4, 7–8, 10, 73, 77, 122, 127, 145, 148, 151, 161, 186n33, 200n34; frontline

work, 14–15, 151; and intentionality, 18; justice blockers, contingent on, 18, 21; motivation and obstruction, tension between, 18; persistence of, 153; as rendered inevitable, 18
inequity, 18, 186n31, 186n33; racial, 158
injustice, 3, 10, 15, 17, 97, 117–18, 144, 146–47, 150–52, 159; perception of, 200n34
intentionality, 18
International Building Code (IBC), 193n37
Intuitionist, The (Whitehead), 97
Iroquois Theater, 61

justice blockers, 3, 13, 17, 21, 149, 151, 153, 161; building court, 156; frontline workers, 145–46, 152; prevention of justice, 12; property markers, 137; property owners, 134–44; rental market, 134; stabs at justice, 130, 143–45, 147; stacked deck, exacerbating of, 143–45; uneven access to resources, 130; unstacking the deck, 130

Kohler-Hausmann, Issa, 140

Lamont, Michèle, 186n37, 192n13, 196n31
landlords, 15, 41–42, 45–46, 76–77, 124, 153, 191n97; bad, 78–80, 96; and building inspectors, 94, 133; corporate, 190n73; inspection reports, ignoring of, 130; judges, siding with, 131; "landlord hell," 20, 81, 132; landlordism, 40, 89–90; negative connotations of, 89; negligence of, 2, 79–80, 132; professional, 60, 83–85; and profit, 80, 86, 89, 92, 133; as rentiers, 89; repairs, and rent hikes, 132–33; retaliation of, 94; small-time, 52, 60, 86–90, 93, 196n18; stacked deck, 86, 133. *See also* property owners
land use laws, 61
limited-liability corporations (LLCs), 43, 92, 190n81, 200n28
London: Grenfell Tower fire, 148, 160; housing stock, 148

malign neglect, 83, 89. *See also* negligence
marginality: and privilege, 4
Massey, Doug, 123
McCall, Leslie, 122
Miami (Florida), 72; condo collapse, 159–60
milking, 29, 41

mom and pop buildings, 86–87. *See also* small-time landlords
Moore, Natalie, 24–25, 36
Morgan, Kimberly J., 187n49
mortgage securitization, 158, 188n19, 190n81
Mueller, Jennifer C., 198n22
Municipal Drug and Gang House Enforcement Program, 190n90

Nagle, Robin, 168
negligence, 16, 78, 160; corner-cutting, 82; and corruption, 116; and greed, 120; of landlords, 2, 79–80, 132; negative connotations of, 89; and profit, 17, 41, 55, 77; and slumlordism, 41; willful, 41
neighborhood change, 11, 83, 106, 110, 112, 150. *See also* changing neighborhoods
neighborhood old-timers, 106–10, 112, 126
New Deal, 28
New Orleans (Louisiana), 72, 148; Hard Rock Hotel collapse, 160

Obama Presidential Center, 41, 197n10
Office of the Inspector General, 176
Operation Crooked Code, 65
Orloff, Ann Shola, 187n49
Our Lady of Angels School Fire, 61
overcrowding, 61, 199n14

Pattillo, Mary, 7, 23, 186n20
policing, 15, 43, 56, 145; broken windows–style, 6, 77; disorder, perceptions of, 8; and leniency, 97; petty crime, overpolicing of, 12; racial disparities in, 10
poverty, 3, 10, 16, 77, 81, 125, 133, 150, 154, 169, 192n12; among Black communities, 121; in communities of color, 124; culture of, 119, 122; and dilapidation, 19, 78; evictions, 45; profiting from, 54; racialized, 17, 147; renters, 132; on South Side, 29, 95; on Southwest Side, 29; stabs at justice, 13; structural racism, 121; urban, 124; vacant lots, 20, 79; and wealth, 4, 6, 22–23, 47, 72
precarity, 72, 97, 100; changing neighborhoods, 110, 112; and greed, 119; and profit, 6–7, 16, 21, 76, 110, 116–18, 126, 152
predatory inclusion, 29–30
predatory lending, 22, 53, 79, 120
procedural hassles, 140–42

profit, 22, 29–30, 38, 40, 44, 66, 83, 85, 113, 119, 150, 157–58, 193n35; affordable housing, 154; corner-cutting, 78; as disproportionate, 81; disrepair, 79; and greed, 112; and landlords, 80, 86, 89, 92, 133; and negligence, 17, 41, 55, 77; and poverty, 54; and precarity, 6–7, 16, 21, 76, 110, 116–18, 126, 152; proportion, 78, 81, 84; rent gaps, exploiting of, 39; rentiers, 81; and repairs, 161; shortcuts, 82; 311 calls, 110; undeserved, 112; as unfair, 80, 86, 89, 96, 114, 117, 126, 192n14; unregulated, 153; zombie properties, 119
property markers, 140; building violation records, 134–35; detrimental effects of, 134–37; home ownership, 137, 143; as justice blockers, 137
property owners, viii, 2–3, 15, 17, 19–21, 28–29, 32, 34–35, 43, 49, 52–53, 55, 59, 61, 67, 69–70, 72, 79, 86, 88, 90, 92–93, 99, 101, 104, 112, 114, 129–30, 141, 150–51, 153, 159, 161, 173–75, 196n18; code enforcement, 94, 134; of color, 124, 126, 143, 188–89n37; demolition, liable for, 144; foreclosure, role of, 156–57; inequity among, 146; leniency and compassion toward, 68, 102–3, 105, 119, 134, 136, 172; property markers, 137, 140; protection of, 100, 155; repairs, 138–39, 142–43, 154, 156–57, 200n31; "trying to do the right thing," 102. *See also* landlords
public housing: demolition of, 41; opposition to, 109
Purifoye, Gwendolyn, 5

racism, 57, 161, 172, 192–93n19, 198n22; eviction, threat of, 131; racial capitalism, 31–32; racial covenants, 28; racial discrimination, 124; racial inequality, 17, 72; racial profiling, 15, 110; racial wealth gap, 6
rebranding, 35
receivership, 39, 112, 114, 141; grants, 157; house sitting, 113; limited, 189n68; receiver agencies, 189–90n69
redlining, 5–6, 22, 28–29, 32, 34–36, 74, 120
rehabbing, 34–35, 40, 45, 82, 95, 118
relative assessments, 20, 77–78, 90, 93, 98
rental buildings, 38; affordable housing, 154, 201n9; big buildings, 15, 84–86; building code violations, 76, 100, 133; "deconverted" to, 39; LLCs, owned by, 43; negligence, 78; rehabbed, 37; size of, 89; small buildings, 15, 19, 84, 88–89, 93, 154, 201n9
rent control, 153–54
renters, 7, 21, 37, 45, 59, 88, 93–94, 98, 130, 148, 153–54, 159, 161, 196n29, 201n9; affordable housing, 133; as Black, 46, 133, 191n104; of color, 95, 131; as Latinx, 46, 133, 191n104; poverty line, 132; warranty of habitability, 131; as White, 46, 107, 131
rent gaps, 31
rent hikes, 7, 130, 132–33
rentiers, 31, 46, 89
rent regulations, 153
rent strikes, 44
reparations, 10, 157–58
reparative justice, 15; toward homeowners, 16
Restorative Housing Program, 157–58
retribution, 10
retributive justice: toward landlords, 16
reverse filtering, 45
Rosen, Eva, 41–42
Rotella, Carlo, 36

Sabbeth, Kathryn, 196n29
safety nets, 159; absence of, 158
Satter, Beryl, 28–29, 168
scales of investment, 110
Scott, James, 11, 151
Section 8, 41, 85–86; housing vouchers, 96
segregation, 23–25, 32, 79, 188–89n37; economic, 103; racial, 72, 109; residential, 15
Seligman, Amanda, 197n10
Shedd, Carla, 5, 186n31
situated bureaucrats, 54
slums: clearance of, 28; slumlordism, 3, 22, 41, 81
small-time landlords, 52, 60, 86–89, 93; affordable housing, 90. *See also* mom and pop buildings
smart buildings, 152
social locations, 8–9, 18, 54–55, 73
Solnit, Rebecca, 7
stabs at justice, 3, 10–12, 16, 18, 126; building inspectors, 17, 20–21, 53, 74, 80, 88–89, 93–94, 97–98, 100–101, 107, 114, 116, 125, 127, 130, 135, 137, 139, 141, 144, 150–52; court inspectors, 129–30, 140, 143–44; discretionary, 152; frontline workers, 97–98, 144, 146; injustice, turned into, 152; justice blockers, 130, 143–45, 147; and leniency,

20; obstruction of, 130, 145; and poverty, 13; and punishment, 20; rental housing, 97–98; stacked deck, 13, 21, 101, 130, 147; unregulated rental market, 129–30
stacked decks, 9, 19–20, 25, 33, 153; affordable housing, 133; and banks, 6; building inspectors, 16, 45–47, 52–54, 58–60, 77, 93–94, 97, 112, 120, 126–27, 133; as call to action, 8; cities as, 18, 22; communities of color, 120; consistency of, 7; court inspectors, 129–30; criminal justice, 5; different aspects of, 10; in educational settings, 5; frontline workers, 13–14, 72; and geography, 36; hierarchies of institutionalized inequality, 149–50; homeownership, 155; housing conditions, 5, 16, 47, 49; housing landscape, 16; injustice frames, 186n24; institutional contexts, 12; invisible elbow, 185n5; justice blockers, 143–45; landlords, 86, 133; low-income minority renters, 46; motivated by, 8, 10–12, 150, 161; as motivating and obstructing stabs at justice, 21; perceptions of, 52, 54, 58, 77, 93, 126–27; persistence of, 13, 130, 146–47; public transit, 5; racialized hierarchy of, 137; racialized landscape of, 119–20; racism, steeped in, 31; as rallying cry, 150; regressive property tax assessments, 5; as relational concept, 4, 7, 17; rentiers, in favor of, 46; resignation, feelings of, 145; small-time landlords, 86–87; social location, 54; stabs at justice, 13, 101, 129–30, 147; taxation, 5; tenants, 97, 191n97; as tool, 52; White neighborhoods, 36; Whites, benefited by, 54, 119, 124, 158
Strategic Task Force (STF), 43–44, 90–93, 190n90, 202n1
straw buyers, 39
structural inequality, 3, 126
subprime loans, 30
subsidized housing, 41
Summers, Nicole, 154–55

tax system, 6
Taylor, Keeanga-Yamahtta, 6, 158
tenants, 6, 77, 93, 95–96, 130, 134, 154; crime, association with, 94; displacement, 3; housing conditions, 3; lower property values, association with, 94; neighborhood collective efficacy, decrease of, 94; presumptions about, 94; relational aspect, 17; rent hikes, 132–33; rent withholding, 190n95; repairs, input of, 155; slumlordism, 81; stacked deck, 97, 191n97; tenants' organizations, 153; threat of eviction, 131; 311 requests, 132, 177; warranty of habitability, 131
311 requests, 19, 53, 58, 64, 66, 76, 95–96, 99–100, 103, 106, 125, 128, 132, 135, 169, 176–77, 191n3, 195n59; building code violations, 49, 52; building conditions, 49; communities of color, 198n29; prioritizing of, 194n45; and profit, 110; property markers, 136
311 system, 194n46; Chicago Buildings Department, funneled through, 64–65; institution of, 63
Tilly, Charles, 185n5
Tonkiss, Fran, 58

United States, 59, 72, 94, 106, 123, 146, 148–49, 155, 158; homeownership, valorization of, 100; home repairs, assisting with, 201n16; house fires, 187n45
unregulated rental market, 129; stabs at justice, obstruction of, 130
upscaling, 3, 22
urban governance: front line of, 13, 17
urban renewal, 42

vacant buildings, 14, 31, 148
Valverde, Mariana, 61, 187n49
virtuous marginality, 108–9

warranty of habitability, 131, 154, 198n2
Water Department, 192–93n19
Watkins-Hayes, Celeste, 12, 54, 187n50
wealth extraction, 22, 32, 100, 124; on Southwest Side, 53; Whites, profiting from, 122–23
welfare reform, 145
White blind spots, 17, 52, 59, 122–23, 125, 158, 197n21, 198n28; vs. colorblindness, 124; disinvestment, 124; racial domination and relational character, obscuring of, 124
White flight, 6, 28, 32, 36, 112
Whitehead, Colson, 97
White neighborhoods: prioritizing of, over communities of color, 35–36; privileges of, 35; stacked deck, integral piece of, 36; urban policies, product of, 35

Whiteness, 35, 59; as construct, 124
White privilege, 59, 197n21
Whites, 31–32, 188–89n37; as colorblind, 198n22; homeownership and wealth, 59; stacked deck, in their favor, 124, 158
White wealth, 34, 36, 151; as hidden, 35; as inconspicuous, 135; as normalized, 124; racist views, expression of, 171

Wilson, William Julius, 29–30
Woodlawn Housing Preservation Ordinance, 157

Young, Iris, 18

zombie properties, 31, 118, 188n30; and profit, 119
zoning, 34, 74; exclusionary, 5–6

Lightning Source UK Ltd.
Milton Keynes UK
UKHW010003201022
410757UK00003B/47